COASTAL FISHES
OF THE PACIFIC NORTHWEST

c·o·a·s·t·a·l FISHES

OF THE PACIFIC NORTHWEST

BY ANDY LAMB AND PHIL EDGELL

Preface by Murray Newman, Ph.D.

HARBOUR PUBLISHING
1986

DEDICATION

Coastal Fishes of the Pacific Northwest is dedicated to all anglers, divers, commercial fishermen, seafood gourmets and naturalists who share our fascination with fishes and to several very special enthusiasts who inspired us to write this volume.

Anglers: Alec Lamb, John Rawle
Divers: Lou Lehmann, Bernard P. Hanby
Beachcombers: Roy Edgell, Charlie Moffett
Commercial Fishermen: the Albert Radil family
Seafood Gourmets: Lauralee Edgell, Virginia Lamb

COASTAL FISHES OF THE PACIFIC NORTHWEST
Copyright © 1986 Andrew Lamb and Philip Edgell

Published by Harbour Publishing Co. Ltd.,
PO Box 219, Madeira Park, BC, Canada V0N 2H0.

Cover and interior design: Gaye Hammond
Back cover photo: Virginia Lamb
Printed and bound in Hong Kong by Colorcraft Ltd.

CANADIAN CATALOGUING IN PUBLICATION DATA

Lamb, Andrew, 1947-
  Coastal fishes of the Pacific Northwest

  Includes index.
  Bibliography: p.
  ISBN 0-920080-75-8

  1. Fishes-Northwest Coast of North America-
Identification.   2. Fishes-Northwest coast of
North America-Pictorial works.   I. Edgell,
Philip.   II. Title.

QL623.4.L35 1986   597.092'63   C85-091486-8

TABLE OF CONTENTS

PREFACE

It gives me great pleasure to have the opportunity to write a brief preface to this work, as I have known both authors since their first involvement at the Vancouver Public Aquarium as young aquarists many years ago.

If you are a fisherman, a diver, or a naturalist, this is the book for you. Andy Lamb has succeeded beautifully in his stated purpose of bridging the gap between scientist and layman. Entertainingly readable, the text succeeds in providing easy and accurate information which will neither raise objection from professionals nor confuse amateurs; it is enhanced by his clear, accurate and beautiful pencil drawings. Mr. Lamb has truly achieved the happy medium. The categorization of the text according to people's different interests is consistent throughout and conveniently usable.

Andy Lamb has a true love of the subject and *Coastal Fishes of the Pacific Northwest* is a product of over eighteen years of devotion to diving and collecting along the coasts of Washington and British Columbia. He is very knowledgeable about where to find particular fish species and his record keeping is excellent. There is in this book a wealth of original observations which have not been published elsewhere, reflecting both his field experience and academic training.

Phil Edgell is one of the top underwater photographers on the BC coast, and a great SCUBA diver (both Phil and Andy have been described by their friends as 'divaholics"); he was one of the first to photograph the beautiful candelabra coral along the BC coast. Superbly skilled, he works extremely hard to find the fish in its natural setting so the viewer of the photograph becomes aware of the nature of the habitat. The fish you see in this book are in their natural environments—not studio settings—so the pictures contribute to the educational value of the work.

Together they have created a unique book which will be indispensable to anyone who is interested in the world lying beneath the surface of our Pacific Northwest coastal waters.

Murray A. Newman, Ph.D.
Director, Vancouver Public Aquarium

ACKNOWLEDGEMENTS

Without the very special efforts of many people we would have been unable to produce this book and we would like to thank them all, but that is almost impossible.

Collection and transportation of living specimens for the photography and artwork was a major and time-consuming requirement. The following people diligently assisted with this task: Mark Ahern, John Allan, George Baker, Joe Bauer, Dr. Richard Beamish, Jim Cave, Gord Cox, Ted Davies, Dr. John Davis, Ralph Delisle, Jack Dibold, Lauralee Edgell, John Gawron, Duane Goertson, John Goertson, Martha Goertson, Martha Goertson Jr., Rita Goertson, Roger Goertson, Mike Gray, Dr. Ian Haas, Barbara Hale, Bernie Hanby, Bob Harrison, Coleen Hay, Ian Hay, Jack Heading, Gil Hewlett, Colin Jerome, Ernie Jerome, Greg Jerome, Mary Jerome, Abby Jones, Dick Jones, Dr. George Kruzinski, Virginia Lamb, Finn Larsen, Don Lawseth, Joyce Lockhart, Neil McDaniel, Dr. Don McPhail, Don McQuarrie, Brian Markert, Jack Markert, Susan Markert, Sandy Matheson, Dr. Charles Moffet, Jenny Moffet, Mimi Moffet, Sally Moffet, Sarah Moffet, Garnet Rawle, John Rawle, Ian Shand, Boyd Shirley, Edith Shirley, Paul Shirley, Dr. Susan Shirley, Andrea Smith, Dee Smith, Judy Smith, Dr. John Stockner, Jim Thompson, Les Tulloch, Jergen Westrhiem, Dr. Norman J. Wilimovsky, and Billy Wong.

Other people contributed photographic, artistic and literary skills: Helen Dye, Roy Edgell, Dr. Larry Giovando, Jack Grundle, Bernie Hanby, Sue Hanby, Bob Hodge, Bess Lamb, Lou Lehmann, Linda Manheim, Dr. Jeff Marliave, Dr. Murray Newman, Dr. Chris Pharo, Gordon Soules, Peter Vassilopoulis.

The following institutions and businesses generously contributed their facilities: Bamfield Marine Station, *Diver*, Harrison's Fishing Lodge, Pacific Biological Station, Pacific Bio-Marine Laboratories, West Vancouver Laboratory, University of British Columbia, University of Washington, Vancouver Public Aquarium, *Western Angling*, Westport Aquarium, Wordtex.

Scientific editing was graciously and thoroughly completed by Dr. Jeff Marliave, Research Scientist at the Vancouver Public Aquarium.

Finally, this book would not have been possible without the careful, painstaking editing of Gordon Elliott. His many hours of consultation with the senior author not only created a polished work from a rough draft but produced a more mature and confident writer.

INTRODUCTION

When people first began to study the fishes along the Pacific shores of North America many amateur naturalists worked side by side with professional fish biologists, and together they slowly and steadily developed an understanding of the fishes of the region. However, as technology increased, so unfortunately did the gulf between career ichthyologists and curious lay people. Technical jargon, complex mathematical interpretations, and computer wizardry have now essentially isolated the many interested amateurs from the sophisticated professionals. Ironically, other impressive technical developments now provide lay naturalists with more leisure time in which to observe fishes by using further modern advances such as SCUBA and photography. While the two groups should have been coming together, the gap has steadily widened.

Coastal Fishes of the Pacific Northwest is an attempt to bridge the gap between the scientist-biologists and the laymen, and to help anyone to identify a fish quickly and accurately, but the book uses a minimum of technical terminology. It classifies special information usable by anglers, divers, beachcombers, commercial fishermen, dockside observers, and seafood fanciers, or by anyone with an interest in marine biology.

Within this volume, closely related species of fish appear together in their respective families and a family discussion page prefaces each grouping. Each family discussion page generalizes about the particular family, presenting biological and ecological information about it, including its definition, its world-wide distribution, its size, its largest member, its history, its importance to man, and its distinctive features, as illustrated by a labelled line drawing. Order of appearance of each species after its relevant family discussion page is designed to arrange the most similar fish together, essentially forming, through the photographs and labelled shaded drawings, a naturally progressing visual key.

In looking at each species, first we consider names. The "common name," the name most people use, denotes the correct English-language term assigned to each fish, the term designated by an international committee of ichthyologists. The "scientific" name, headed SPECIES, which is usually given in Greek or Latin, allows for precise identification anywhere and in any language, and indicates relationships much as do human surnames. Not only does "Bill Smith" define a specific person, it also implies a relationship with other "Smiths". This bi-nomial, or two-word, scientific name for a fish demonstrates the same principle, except that in the scientific biological name the first word is the "surname". Consider, for example, the copper rockfish *Sebastes caurinus: Sebastes* corresponds to "Smith," while *caurinus* corresponds to "Bill". *Sebastes caurinus* is, therefore, closely related to the many other *Sebastes*, or rockfish species. Do not be afraid of those technical names, and do not be afraid to say them: pronunciation generally follows standard rules for English. In fact, one need not learn the technical terms at all, but familiarity with the scientific name does allow access to the entire literature on that fish.

The ALTERNATE NAMES can cause confusion because, for cultural, geographic or economic reasons, one species might have more than one name. More confusing is that two different fishes can have the same name. Alternates listed here are in order of acceptance, and those marked with an asterisk are totally unacceptable; users of the unacceptable should change to the accepted common name.

MAXIMUM RECORDED SIZE and DISTRIBUTION are of less importance, perhaps, than name and scientific name but still are of interest. The maximum recorded size is exactly that, the size recorded by responsible documentation. Unfortunately the measurements as stated do not always represent their size categories with classic precision. Some species, particularly those that are either economically significant or smaller in size, have historically received more intense statistical documentation. Consequently figures presented serve as the best values available and are often rounded off to the nearest appropriate size. Anyone finding a specimen larger than that recorded here should report that finding. Measures used are metric with equivalents in parentheses. Distribution also involves recording. The outside limits given indicate documented extremes of the species' range; anyone finding the species beyond these limits should document such a discovery. Send reports on both size and distribution to the authors of this book who will distribute that information.

The illustrations of each fish, the photographs and the sketches, show some variations in the species. The colour photograph is usually of a typical adult depicted in a lateral, or "side on" pose, and offers the colours necessary for positive identification. The fish were photographed by Phil Edgell "on location," or were captured and then released in appropriate habitats as close to natural as possible. For the shaded, labelled, and proportionately accurate sketches Andy Lamb used live healthy specimens as models. Occasionally, when a species was too big to hold captive, such as the sixgill shark, a composite was sketched from the many colour slides accumulated while taking photos. Shading and proportion give the feel of life and nature. In addition to the usual lateral view, insets sometimes illustrate specific parts of sexual or other variations. Particularly important features are labelled on each drawing, focusing the reader's attention on the structures that make this species distinctive.

Basic to *Coastal Fishes of the Pacific Northwest* are the special passages marked with silhouette symbols easily identified by the angler, diver, beachcomber, commercial fisherman, and seafood fancier, all people with special interests in the fishes. The designated passages tell what, why, where, and when specific fishes may be encountered. Scattered throughout these five special passages are biological facts which come out of specialized scientific research. For example, a scientist might have spent several years learning what a cod feeds on and we summarize for the angler and commercial fisherman what baits to use.

SYMBOLS

The silhouette of the person with a hook and line represents the people primarily interested in sports fishing and the passage accompanying the symbol contains suggestions about techniques, bait and tackle. This book does not make a value judgment about species angled for, because judgments depend on the personal preferences and prejudices of each angler. Neither do we ask why you are interested in a fish of a specific size or species; such information is also personal. Instead we give tips useful for anyone interested in capturing a species. For instance, you might not want to catch a Pacific staghorn sculpin, but if you do want to catch one you can find some hints in this text. On the other hand, you can apply the "how to catch" ideas negatively: to catch the Pacific staghorn sculpin, fish the bottom with bait; to avoid the Pacific staghorn sculpin, use a fast moving lure. Some anglers might not want a shiner perch for pleasure or for food, but those same fishing enthusiasts may use it for bait.

The passage marked by the diver builds on the identification already made through the drawings and the photographs. Actually, SCUBA diving has opened up new opportunities for both fish biologists and amateur naturalists, and our passages include descriptions of habitat in order to lead even more divers to the fish. Fish behaviour is also of importance to a diver who may have a first hand opportunity to observe such fascinating action. The book encourages exploration to about 36 metres (120 feet) and gives suggestions on what to do and see within practical limits of a diver's ability. Ecology-conscious spearfishermen can use information not only on behaviour and habitat, but also specific information on size, legal limits, speed and movement. Underwater photographers can also use most of this material.

This book does not tell commercial fishermen how to fish, but the section marked with the seiner silhouette is here for those interested in species identification. The information on ecology and behaviour might reinforce the understanding commercial fishermen have already, or stimulate their thinking by raising some questions about their own experience. The layman interested in commercial harvesting can at least learn something of methods, gear, seasonality, and bait.

The strolling beachcomber and wharf-bound observer or curious boat passenger all have terrestrial vantage points from which to view an amazing array of marine life. We offer the beachcomber hints on what he or she might find in tidepools and how to uncover the life under the weeds and rocks exposed at low tide. Water, height and angle hinder the specific identification for dockside observers, but a few tips allow educated guesswork. Boat passengers also gain in knowledge; for example, shorebound naturalists will not be able to see the juvenile tiger rockfish, but those in a boat do have an opportunity. We stress conservation and ask that all observers respect the environment: return captured fish for which you have no use, replace upturned rocks to their original position, and do not litter.

Unfortunately, long-held cultural prejudices still linger about the edibility of certain species. The section with the plate and cutlery suggests going beyond your own cultural biases and experimenting with different fish as food. Much of the marine life of the Pacific Northwest can be interchangeable as ingredients in many fish recipes. Try gunnels in a bouillabaisse for instance. Or try fried tube snouts. You can do anything with dogfish if you know how to treat it. The native Indians and the Japanese have known for centuries how to use herring and kelp. Try herring roe fried in eulachon oil.

Coastal Fishes of the Pacific Northwest is based on scientific research, personal experience, and a belief in conservation. The public should not only enjoy fish more knowledgeably, but could also assist in further research. For these reasons we suggest that people having trouble with identification call on government fisheries facilities, particularly those involved in research, universities with ichthyological departments, large public aquariums, or the natural history section of public museums. If frustrated amateur researchers gain no satisfaction from those places, they should not give up in bewilderment, but should instead write to the authors of this book.

Andy Lamb and Phil Edgell

PHOTO SECTION

Red Irish Lord

SUGGESTED STEPS TO FISH IDENTIFICATION
USING *Coastal Fishes of the Pacific Northwest*

1. If unfamiliar with the various fish families present in the Pacific Northwest, first consult the table of contents and examine the silhouette that accompanies each family listing. Further detail, by way of larger labelled line drawings, is available on the corresponding family pages.

2. Targeting on the family or families that you believe to be appropriate, look through the following colour photographs until you find one or more fish that could be the one you are trying to identify. Those species that are most similar are grouped together, so you have probably reduced the number of possibilities considerably by now.

3. Using the fish number listed with each photograph, find the appropriate page(s) in the text—the fish number is at the top of each text page, in front of the name. Study the labelled shaded drawing(s), together with the corresponding colour photograph(s) to complete the identification. **Be aware there is often variation, particularly in both colour and pattern, between individuals of the same species.** Therefore it is most important to study the significant features labelled on the shaded drawing(s) as well as the colour photograph(s) when comparing the specimen(s) (or photographs) that you are attempting to identify.

4. Within the text for each species, the information presented there, such as distribution, maximum size and natural history details, may be used to help verify the identification you have made.

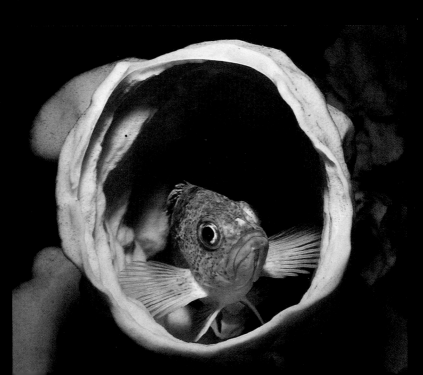

1 Pacific Hagfish

2 Pacific Lamprey

3 River Lamprey

4 Sixgill Shark

5 Brown Cat Shark

6 Spiny Dogfish

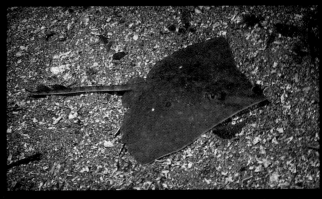

7 Longnose Skate

8 Big Skate

9 Sandpaper Skate

10 Ratfish

11 White Sturgeon

12 Pacific Herring

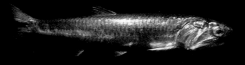

13 Northern Anchovy

14 Eulachon

15 Surf Smelt

16 Capelin

17 Longfin Smelt

18 Pacific Sand Lance

19 Pink Salmon

20 Chum Salmon

21 Sockeye Salmon

22 Chinook Salmon

23 Coho Salmon

24 Steelhead Trout

25 Coastal Cutthroat Trout

26 Dolly Varden

27 Pacific Cod

28 Pacific Tomcod

29 Walleye Pollock

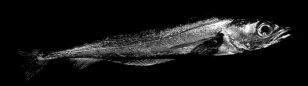

30 Pacific Hake

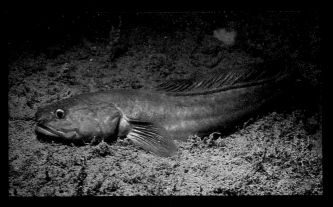

31 Red Brotula

32 Blackbelly Eelpout

33 Black Eelpout

34 Shortfin Eelpout

**35 Threespine
Stickleback**

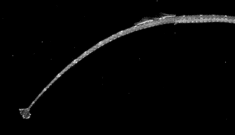

37 Bay Pipefish

38 Pacific Sandfish

39 Striped Seaperch

40 Pile Perch

41 White Seaperch

42 Silver Surfperch

43 Redtail Surfperch

44 Kelp Perch

45 Shiner Perch

46 Blackeye Goby

47 Bay Goby

48 Arrow Goby

49 Northern Ronquil

50 Striped Kelpfish

51 Crevice Kelpfish

**52 Mosshead
      Warbonnet**

53 Decorated Warbonnet

54 High Cockscomb

55 Slender Cockscomb

56 Whitebarred Prickleback

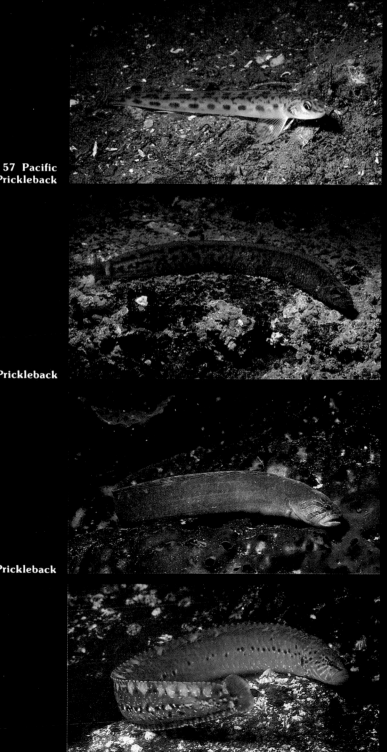

57 Pacific Snake Prickleback

58 Rock Prickleback

59 Black Prickleback

60 Ribbon Prickleback

**61 Longsnout
Prickleback**

**62 Penpoint
Gunnel**

**63 Rockweed
Gunnel**

**64 Crescent
Gunnel**

65 Saddleback Gunnel

66 Longfin Gunnel

67a Wolf-Eel Juvenile

67b Wolf-Eel Adult

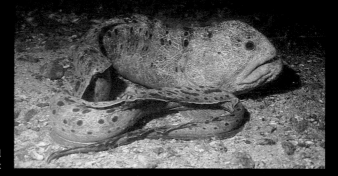

68 Giant Wrymouth

69 Dwarf Wrymouth

70 Quillfish

71 Graveldiver

72 Copper Rockfish

73 Quillback Rockfish

74 Brown Rockfish

75 China Rockfish

76 Yellowtail Rockfish

77 Black Rockfish

78 Blue Rockfish

79 Dusky Rockfish

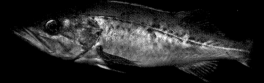

80 Bocaccio

81 Silvergray Rockfish

82 Widow Rockfish

83 Canary Rockfish

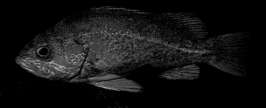

**84 Vermilion
Rockfish**

85 Tiger Rockfish

**86a Yelloweye
Rockfish
Juvenile**

**86b Yelloweye
Rockfish
Sub-Adult**

**86c Yelloweye
Rockfish
Adult**

87 Splitnose Rockfish

88 Puget Sound Rockfish

89 Redstripe Rockfish

**90 Greenstripe
Rockfish**

**91 Shortspine
Thornyhead**

92 Sablefish

**93a Kelp Greenling
(female)**

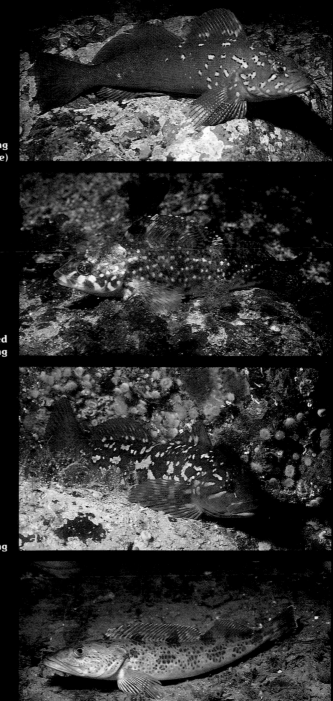

**93b Kelp Greenling
(male)**

**94 Whitespotted
Greenling**

95 Rock Greenling

96 Lingcod

97 Painted Greenling

98 Smoothhead Sculpin

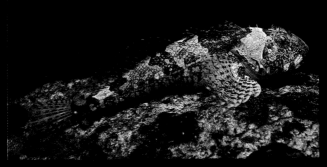

99 Padded Sculpin

100 Scalyhead Sculpin

101 Tidepool Sculpin

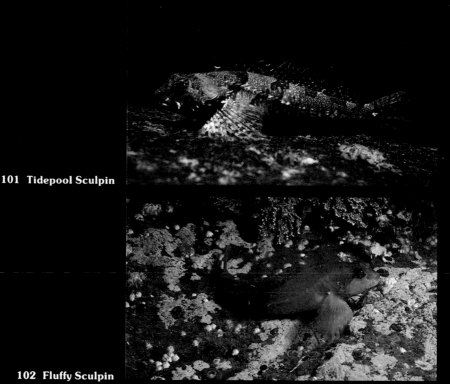

102 Fluffy Sculpin

103 Saddleback Sculpin

104 Sharpnose Sculpin

105 Calico Sculpin

106 Mosshead Sculpin

107 Rosylip Sculpin

108 Prickly Sculpin

109 Manacled Sculpin

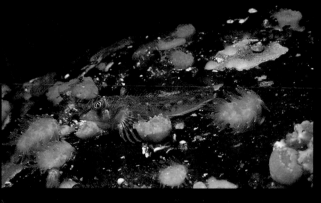

110 Longfin Sculpin

111 Spinynose Sculpin

112 Northern Sculpin

113 Dusky Sculpin

114 Spotfin Sculpin

115 Threadfin Sculpin

116 Ribbed Sculpin

117 Roughspine Sculpin

118 Slim Sculpin

119 Darter Sculpin

120 Roughback Sculpin

121 Spinyhead Sculpin

122 Blackfin Sculpin

123 Tadpole Sculpin

124 Soft Sculpin

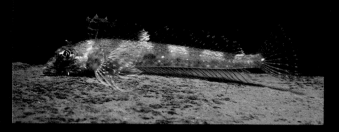

**125 Thornback
Sculpin**

**126 Silverspotted
Sculpin**

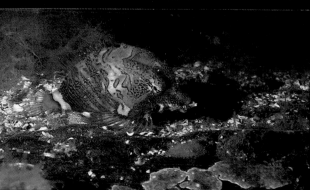

127 Sailfin Sculpin

128 Grunt Sculpin

129 Cabezon

130 Great Sculpin

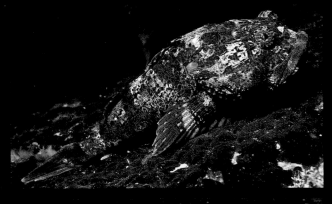

131 Buffalo Sculpin

132 Red Irish Lord

133 Brown Irish Lord

134 Pacific Staghorn Sculpin

135 Plainfin Midshipman

136 Northern Spearnose Poacher

137 Sturgeon Poacher

138 Pygmy Poacher

**139 Spinycheek
Starsnout**

140 Gray Starsnout

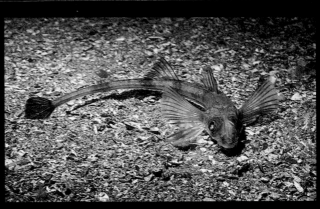

141 Bigeye Poacher

142 Blackfin Poacher

143 Blacktip Poacher

144 Tubenose
Poacher

145 Smooth Alligatorfish

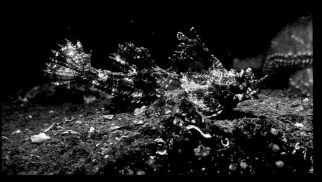

146 Kelp Poacher

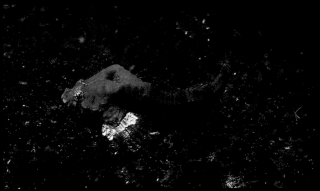

147 Rockhead

148 Pacific Spiny Lumpsucker

149 Tidepool Snailfish

150 Spotted Snailfish

151 Slipskin Snailfish

**152a Marbled
Snailfish**

152b Marbled Snailfish

153 Lobefin Snailfish

154 Showy Snailfish

155a What Species Is It?

**155b What Species
Is It?**

**156 Undescribed
Snailfish**

157 Kelp Clingfish

**158 Northern
Clingfish**

159 Speckled
Sanddab

160 Pacific Sanddab

161 Pacific Halibut

162 Arrowtooth
Flounder

163 Sand Sole

164 Starry Flounder

165 Rock Sole

166 English Sole

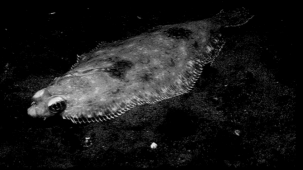

167 Dover Sole

168 Butter Sole

169 Flathead Sole

170 Petrale Sole

171 Slender Sole

172 Rex Sole

173 C-O Sole

174 Curlfin Sole

THE HAGFISHES
(Family: Myxinidae)

The hagfishes, scientifically known as Myxinidae, comprise a small family of fishes containing perhaps thirty known living species which dwell only in the marine environment. Sandy or muddy substrates of the continental shelf are prime myxinid fish domain. Most hagfishes dwell in deep, largely inaccessible haunts. One of two Pacific Northwest species, the black hagfish is not included in this book because it lives in such deep water that people seldom see it.

These blind, elongate creatures, the largest of which may attain 80 cm (32 inches) are among only a few fishes that truly qualify as scavengers. Using their keen sense of smell, myxinids seek out dead or dying fishes that lie on the sea floor. After burrowing inside the bodies of such immobilized prey, these lethargic hagfishes reduce the carcasses literally to skin and bone. If disturbed, particularly while feeding, myxinids may produce amazing amounts of thick, viscous, toxic slime that adheres to virtually anything.

NO EYES

MOUTH WITH EVERTABLE TONGUE
AND 2 ROWS OF TEETH

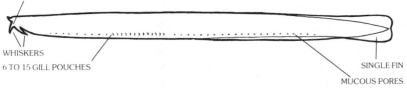

WHISKERS

6 TO 15 GILL POUCHES

SINGLE FIN

MUCOUS PORES

Because of their feeding habits, their ability to produce that slime, and their rubbery and inedible flesh, hagfishes are a bane of fishermen. These greatly scorned beasts, dubbed "slime eels" by many, invade commercial nets in many locales, and with their messy activities, spoil valuable catches. Numerous scientists however, fascinated by the primitive bodily functions of these fishes, employ myxinids as means for further understanding human biology.

The soft-bodied hagfishes possess few hard parts and therefore no fossilized remains of their prehistoric relatives have yet been unearthed. Consequently, the origin of myxinids remains unknown and subject to speculation. Among present-day fishes, though, the scaleless, jawless tubular hagfishes—lacking paired fins—relate most closely to the somewhat similar lampreys.

1

1 Pacific Hagfish

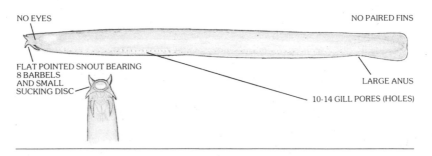

SPECIES: *Eptatretus stouti*—from the Greek *hepta* and *tretos*, meaning "seven" and "apertures;" and honouring Dr. Arthur B. Stout, former secretary of the California Academy of Sciences.

ALTERNATE NAMES: California hagfish, common hagfish, hagfish, hag-fish, hag, slime eel*.

MAXIMUM RECORDED SIZE: 63.5 cm (25 inches).

DISTRIBUTION: Bahia San Pablo, central Baja California, Mexico, to southeastern Alaska.

 To catch a Pacific hagfish with conventional angling gear is virtually impossible, but if capture seems desirable try an effective trap pioneered by research scientists. Repeatedly puncture all sides of a metal can from the outside; this allows specimens to enter the trap easily, but severely hinders their exit. After baiting this trap with fish and securing it with rope, lower it to a muddy bottom at a depth of between 18 and 1,000 m (60 to 3,300 feet). And wait.

 A diver venturing over silty bottoms at depths greater than 18 m (60 feet) might occasionally encounter a Pacific hagfish resting coiled in a figure eight or a tight circular pattern.

 Particularly in California, the Pacific hagfish represents a costly nuisance to commercial fishermen who leave nets out many hours when seeking valuable groundfish. Attracted to the trapped dead-or-dying catch, these despised creatures enter the nets, often in great numbers, then reduce the valuable species to skin and bone. After burrowing into the carcasses and devouring all the flesh, enmeshed Pacific hagfishes secrete quantities of slime, fouling the nets of the distraught fishermen.

 The tidepool observer or wharf-bound naturalist never sights the Pacific hagfish because of its deep dwelling tendencies.

The rubbery texture and the fishy and oily taste of the Pacific hagfish discourage any potential gourmets. The slime is also toxic.

*—incorrect

2

THE LAMPREYS
(Family: Petromyzonidae)

Approximately thirty known living species of lampreys form the Petromyzonidae, "the family of stone rollers." All lampreys reside in the cool temperate or frigid boreal waters of the world—tropical waters are inhospitable to these swift, elongate animals. Although some frequent both fresh and saltwater during different life cycle stages, others live their lives entirely in freshwater habitats. No strictly marine petromyzonids exist however because all species ascend rivers or streams to spawn. Within the Pacific Northwest two anadromous species flourish; as adults, they migrate from the sea to freshwater to spawn and upon entering streams build nests by moving small rocks with their mouths.

MOUTH IS DISC-LIKE FUNNEL
WITH HORNY TEETH

DORSAL FIN

PROMINENT EYES

7 GILL POUCHES

NO WHISKERS

Soon after hatching, tiny larval lampreys, termed ammocetes, burrow into muddy or silty river bottoms and remain there for some months, staying inactive but looking completely unlike their parents, while extracting nourishment from the surrounding ooze. Ultimately a metamorphosis occurs and the tiny ammocetes transform into miniature adults, emerge from the mud and, in the case of marine species, make their way seaward to forage actively. Once they locate prey petromyzonids sink their tooth-studded sucking-disc mouths on to the victims' bodies and begin to rasp holes through skin or scales. Although few lampreys attain lengths of 91 cm (36 inches), they attack many varieties of large fishes, even whales. The swift lampreys relate most closely to the sluggish hagfishes because both lack jaws, paired fins and scales, though they have different feeding behaviours.

Early European and North American cultures highly regarded petromyzonids as food and considered them delicacies. Recently, though, the sea lamprey of the Atlantic devastated major Great Lake fisheries, because when the St. Lawrence Seaway opened up the Great Lakes to marine shipping routes it also exposed lake char and whitefish stocks to this voracious predator. Costly eradication programs eventually controlled the unfortunately introduced menace.

3

2 Pacific Lamprey

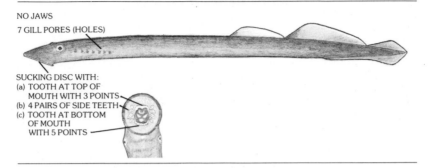

NO JAWS
7 GILL PORES (HOLES)

SUCKING DISC WITH:
(a) TOOTH AT TOP OF
MOUTH WITH 3 POINTS
(b) 4 PAIRS OF SIDE TEETH
(c) TOOTH AT BOTTOM
OF MOUTH
WITH 5 POINTS

SPECIES: *Lampetra tridentatus*—from Latin *lambere*, *petra*, *tri* and *dentatus*, "to suck," "stone," "three" and "tooth."

ALTERNATE NAMES: Pacific sea-lamprey, threetoothed lamprey, tridentate lamprey, sea lamprey.

MAXIMUM RECORDED SIZE: 76 cm (30 inches) and .5 kg (1 pound).

DISTRIBUTION: Punta Canoas, central Baja California, Mexico, to the Bering Sea coasts of Alaska and Asia, then south to the Yuhutu River, Hokkaido northern Japan.

An almost impossible direct catch, the Pacific lamprey may occasionally dangle from a hooked salmon or other gamefish. More likely, though, the angler who catches a salmon, steelhead trout, Pacific hake, sablefish or arrowtooth flounder may notice a circular mark which bears witness to a Pacific lamprey attack. Surprisingly, identical wounds may also pock various whales, including the blue, humpback, sei, finback and sperm. Frequently though, predator becomes prey; sometimes sperm whales feed upon Pacific lamprey.

Very rarely would a diver site the active Pacific lamprey because it dwells at depths inaccessible to sport aquanauts.

Occasionally used for reduction to meal, Pacific lamprey on their spring migration frequently enter the nets of river-based fishermen. Gillnetters and seiners often find some of their harvested salmon carrying either wriggling Pacific lampreys or circular wound scars indicative of past attacks. How many salmonid victims annually die from such attacks is as yet undetermined. Like its anadromous prey the salmon, this lamprey ascends rivers and creeks to spawn. Working together as mated pairs from April to July, Pacific lampreys construct gravel nests, stone by stone with their suctorial mouths.

Shorebound naturalists never sight the Pacific lamprey in shallow accessible marine waters, but dipnet-wielding native Indians and some European immigrants gather specimens in streams.

Some folk eagerly eat fried, broiled, or baked Pacific lamprey while others prefer it smoked.

4

3 River Lamprey

NO JAWS
7 GILL PORES (HOLES)

SUCKING DISC WITH:
(a) TOOTH AT TOP OF MOUTH WITH 2 POINTS
(b) 3 PAIRS OF SIDE TEETH
(c) TOOTH AT BOTTOM
     OF MOUTH
     WITH 6 POINTS

SPECIES: *Lampetra ayresi*—from the Latin *lambere* and *petra*, "to suck" and "stone;" and honouring its discoverer, early naturalist William O. Ayres.

ALTERNATE NAMES: western river lamprey, parasitic river lamprey, western brook lamprey, western lamprey.

MAXIMUM RECORDED SIZE: 31 cm (12.2 inches).

DISTRIBUTION: San Francisco Bay, central California, to Lynn Canal and the Taku River, southeastern Alaska.

Because the Pacific herring is its favourite prey, the river lamprey may attack plug cut herring or herring strip being dragged through the water to lure salmon. Such attacks may be fairly common near the surface from June to September in some locales. If plagued by these predators, retain them as live bait and then bottomfish for lingcod or rockfish, all of which may eat river lamprey.

Divers almost never see the river lamprey unless this predator attacks a herring or small salmon near the surface. Expect only a brief glimpse because these fish avoid the aquanaut, and live in open water.

As an unwanted "bonus" adhering to captured salmon, the economically underutilized river lamprey is occasionally taken by seiners and trollers. Usually this lamprey fastens along the upper side of the host between the victim's head and tail. While larger, stronger salmon often survive these attacks, maintaining circular scars, the smaller Pacific herring is usually completely devoured. Like its anadromous prey the salmon, the river lamprey migrates to fresh water for intricate courtship and spawning activities.

The feeding river lamprey might attack an unsuspecting herring, young salmon, or anchovy adjacent to a wharf or jetty where the naturalist could actually observe the entire activity.

Throughout the world, other cultures feast upon the delicate flesh of lampreys. Skin and try it—any recipe for eels will do. Smoked or fresh, the river lamprey offers potentially fine eating.

5

THE COW SHARKS
(Family: Hexanchidae)

The Hexanchidae, popularly known as the cow sharks, contains only six or seven known species that exist today. Intriguingly though, these few surviving forms range widely, with some, including the Pacific Northwest's sixgill shark, distributed almost worldwide in temperate and tropical regions. Cow sharks live only in the marine environment where, particularly in tropical latitudes, they seek out deep, dark, cool waters. Sluggish creatures, hexanchid sharks swim slowly along the bottom and may rest upon the sea floor.

Reports from natural history expeditions of the 1800s reveal impressive statistics for some members of the Hexanchidae; leviathans of 5.6 m (18 feet) and 771 kg (1,700 pounds). The reproductive capacity of cow sharks is also amazing: a large, prolific female hexanchid shark may individually give birth to as many as 108 fully formed pups at once, each measuring up to 40 cm (16 inches).

Years of paleontological searches have produced an excellent fossil history for the primitive cow shark family. Exhibiting little change throughout the last 150 million years, hexanchid sharks have survived since mid-Jurassic times. Because they possess six or seven pairs of gill slits, instead of the usual five sets prevalent upon most other living sharks, they are believed to be among the most primitive of existing species.

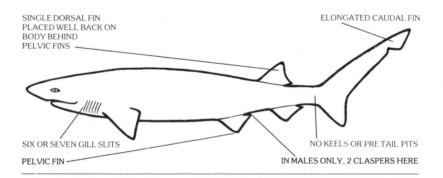

SINGLE DORSAL FIN PLACED WELL BACK ON BODY BEHIND PELVIC FINS

ELONGATED CAUDAL FIN

SIX OR SEVEN GILL SLITS

PELVIC FIN

NO KEELS OR PRE TAIL PITS

IN MALES ONLY, 2 CLASPERS HERE

Not so long ago, modest but steady fisheries took the massive cow sharks in various regions of the world. Improved technology, though, provided synthetic oils that are finer and cheaper than those rendered from the huge hexanchid shark livers and the once lucrative cow shark fisheries faded away. Perhaps some as yet unrecognized attribute of these giant animals will prove beneficial to humanity and rekindle interest in these once profitable enterprises.

One species of Pacific Northwestern hexanchid is the sevengill shark which, although easily recognized by its seven pairs of gill slits and spotted body, lives in deep water and is a most unlikely sighting.

4 Sixgill Shark

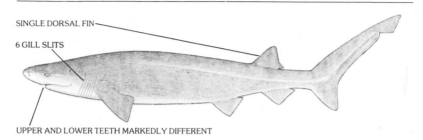

SINGLE DORSAL FIN

6 GILL SLITS

UPPER AND LOWER TEETH MARKEDLY DIFFERENT

SPECIES: *Hexanchus griseus*—from the Greek *hex* and *ankos*, "six" and "bend;" and the Latin *griseus*, "gray."

ALTERNATE NAMES: Six-gilled shark, sixgill cowshark, shovelnose shark, mud shark.

MAXIMUM RECORDED SIZE: 4.7 m (15.5 feet) and 590 kg (1,300 pounds) for a Pacific specimen.

DISTRIBUTION: Throughout the world's temperate seas and along the Pacific coast of North America from Bahia San Todos, northern Baja California, Mexico, to the Gulf of Alaska.

 The giant, sluggish sixgill shark would most likely swim past the small baits inadvertently offered by anglers, but if this shark did seize a hook, the conventional Pacific Northwest sport gear would not subdue the immense body of this fish. The sixgill shark's lethargic, plodding behavior attracts very few sportsmen to employ the heavy tackle and wire leader necessary against this creature's sharp teeth.

Especially in British Columbia, the bulky, slow sixgill shark may drift past wary and often surprised divers. Though usually most abundant below regular SCUBA limits, this gray, green-eyed beast invades shallow, diveable locales primarily during spring and summer. Exercise care, because this generally docile shark could be provoked into aggressive behaviour; its formidable teeth and strong jaws are very dangerous.

 Sizeable liver oil and reduction fisheries for the sixgill shark operated in the Pacific Northwest before the Second World War. The commercial gear consisted of long set lines, each having large hooks baited with dogfish or chunks of meat. Much to the dismay of net fishermen a sixgill shark occasionally blunders into their gear and causes great havoc. Inexpensive, chemically manufactured products replaced sixgill shark liver oil and spelled the demise of its fishery.

 Very rarely an observant boater might notice the broad, large shape of a sixgill shark lurking in shallow, extremely clear water.

 Somewhat popular in some parts of California, sixgill shark flesh is apparently edible!

THE CAT SHARKS
(Family: Scyliorhinidae)

The Scyliorhinidae, the "skin nose family," is popularly known as the cat shark family and of the nearly sixty known living species within this group, few grow to more than a modest size: 122 cm (48 inches). The small, bottom-dwelling cat sharks live throughout the world's temperate and tropical seas, flourishing particularly in the Indo-Pacific region. Indeed some of the most attractively coloured and patterned sharks are IndoPacific scyliorhynids. Most cat sharks inhabit shallow inshore waters swimming sluggishly near the bottom and taking frequent rest intervals. Only a few species populate deep water haunts: one is the brown cat shark, the single family representative in the Pacific Northwest.

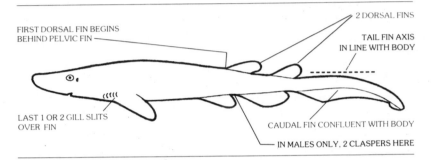

FIRST DORSAL FIN BEGINS
BEHIND PELVIC FIN

2 DORSAL FINS

TAIL FIN AXIS
IN LINE WITH BODY

LAST 1 OR 2 GILL SLITS
OVER FIN

CAUDAL FIN CONFLUENT WITH BODY

IN MALES ONLY, 2 CLASPERS HERE

Several months after mating, during spawning season, a female brown cat shark deposits several horny egg cases containing developing young upon the sea floor. These life-bearing rectangular packets are species-specific in shape and most often have curly, thin tendrils dangling from each corner; these tendrils become entangled among bottom debris and act as anchors for the egg cases. After weeks of developing into miniature editions of scylioryhinid sharks but still within a packet, the young squirm free of their protective surroundings and immediately begin a search for food in their new and larger environment.

Not significantly different from today's versions, prehistoric cat sharks apparently swam throughout shallow seas during the upper Jurassic period. Evidence for this theory stems from scyliorhinid fossils which date back at least 135 million years. Among the vast array of present-day sharks and rays, the slow and bottom dwelling nurse sharks most closely relate to the scyliorhinids. Although these economically insignificant creatures have never been harvested, perhaps a new product will create a demand for the readily available cat sharks.

5 Brown Cat Shark

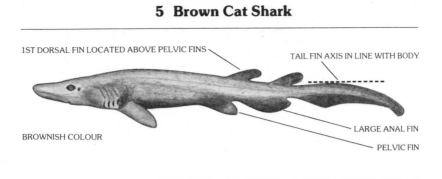

1ST DORSAL FIN LOCATED ABOVE PELVIC FINS

TAIL FIN AXIS IN LINE WITH BODY

BROWNISH COLOUR

LARGE ANAL FIN

PELVIC FIN

SPECIES: *Apristurus brunneus*—from the Greek *a*, *pristis*, and *durus*, meaning "without," "file," and "hard;" and the Latin *brunneus*, "brown."

ALTERNATE NAMES: catshark, brown shark*

MAXIMUM RECORDED SIZE: 68 cm (26.8 inches).

DISTRIBUTION: northern Baja California, Mexico, to Icy Point, southeastern Alaska.

 The flaccid, sluggish brown cat shark with immobile pupils in its eyes, lives in deep water over muddy or sandy bottoms—a region seldom fished by the angler. This rarely hooked shark consumes shrimps and small fishes, which should make good bait for it.

 Considering the brown cat shark's minimum depth record is 33 m (109 feet), the diver should not expect to sight many specimens. Finding this shark's rectangular 5 cm long egg cases with string-like tendrils at each corner, also rates as a noteworthy event. Adult female brown cat sharks jettison these packets between February and August, and the young apparently take a year to develop within their safe confines.

 Shrimp and groundfish trawlers unfortunately take many brown cat sharks when dragging nets over sandy, muddy bottoms at depths to 1,190 m (3,900 feet) and occasionally deep mid-water sweeps capture specimens well off the bottom. Although this soft, flabby species is of no direct economic value, it perhaps falls prey to many of the valuable groundfish.

 Unless washed ashore after a violent storm a brown cat shark egg case or even less likely, a brown cat shark carcass, is never encountered by the beachcomber or dockside naturalist.

 The soft, unappealing flesh of the brown cat shark negates any desire to eat it or to prepare it for eating.

*—incorrect

9

THE DOGFISH SHARKS
(Family: Squalidae)

Within the moderately sized family of sharks known as the Squalidae, a natural division occurs between the spiny dogfishes and the sleeper sharks. Possessing spines at the front of each dorsal fin, the 47 living species of spiny dogfishes form the largest of the two subgroups; the sleeper sharks, with spineless dorsal fins, number only 9. All squalid sharks live in warm or cool seas and the various species are spread rather evenly throughout the temperate latitudes. A great variety of habitats, from inshore shallows to abyssal offshore haunts, harbour the various squalid sharks, though a majority, including some with luminescent organs, swim at great depths. The tiny light sources that dot the bodies of some deepwater dogfish undoubtedly aid social interaction at almost lightless depths.

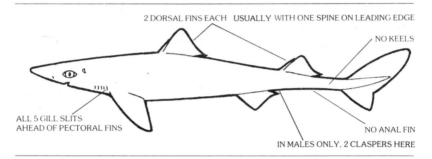

2 DORSAL FINS EACH USUALLY WITH ONE SPINE ON LEADING EDGE

NO KEELS

ALL 5 GILL SLITS
AHEAD OF PECTORAL FINS

NO ANAL FIN

IN MALES ONLY, 2 CLASPERS HERE

Most of the squalid sharks, including all of the "spiny forms," attain only modest maximum sizes, with few growing larger than the 160 cm (63 inches) spiny dogfish so common in the Pacific Northwest. A few giants exist among the "spineless" forms which may stretch to an impressive 7.6 m (25 feet).

A great amount of discovered fossil material assists scientists in tracing the origins and pedigree of dogfish sharks, with inspection of these prehistoric remains indicating the squalid lineage dates back at least 135 million years to the upper Jurassic period. Historical and cultural disparities reflect how mankind views dogfish sharks: in Europe the cosmopolitan temperate spiny dogfish attracts great interest whereas locally few venturesome people consider dining upon it. Future generations will perhaps look more favourably on this locally rejected species.

The only other local squalid shark, the massive Pacific sleeper shark, inhabits such deep water that people seldom see it.

6 Spiny Dogfish

MAY HAVE WHITE SPOTS ON SIDE OF BODY

2 DORSAL FINS, EACH WITH 1 SPINE ON LEADING EDGE

NO ANAL FIN

SPECIES: *Squalus acanthias*—from the Latin *squalus*, meaning "shark;" and the Greek *akanthias*, "spines."

ALTERNATE NAMES: Pacific dogfish, piked dogfish, Pacific grayfish, grayfish, spurdog, mud shark, sand shark*, salmon shark*.

MAXIMUM RECORDED SIZE: 160 cm (63 inches) and 9.1 kg (20 pounds).

DISTRIBUTION: throughout much of the North Atlantic Ocean and Mediterranean Sea, as well as the Pacific from San Martin Island, southern Baja California, Mexico, to the Bering Sea coast of Alaska and the Aleutian chain.

 Much to the chagrin of most bait fishermen, the spiny dogfish commonly bites the hooks directed towards more popular quarry. Ironically, though, some anglers actively seek this shark as a welcome, sporty catch but should use wire leader to prevent its sharp teeth from severing the line. Handle this writhing catch carefully: avoid its snapping jaws and its sharp dorsal fin spines.

 Sporadically the spiny dogfish becomes a schooling fish, and an active companion for the diver. Such encounters are somewhat unnerving because these small but fierce looking animals may quickly dart toward the diver, and then suddenly veer away, showing speed and grace and power.

In addition to its edible flesh, the spiny dogfish has provided sandpaper from its skin, and oil and vitamins from its large liver. Throughout history, however, commercial fishermen have alternately reviled and actively harvested this infamous or famous shark. In attempts to "rid it from the sea" governments have posted bounties; at other times these same bureaucracies have tried to establish viable, continuous fisheries for dogfish.

 The active spiny dogfish frequently swims at the surface, finning or thrashing in pursuit of plankton, herring or other small fishes. Watch for this action from shore, wharf or boat.

 Russ Mohoney's *The Dogfish Cookbook*, an excellent 1976 publication, tells how to prepare and enjoy the surprisingly delicious spiny dogfish.

* —incorrect

THE SKATES
(Family: Rajidae)

The large family of skates, scientifically termed Rajidae, contains slightly more than 100 known living species. Coastal regions of the continental shelves in cool northern or temperate seas harbour most of these broad, thin creatures, including the world's largest, the big skate of the Pacific Northwest. Some rajids though are found in deep dark water, upon the sea floors adjacent to the northern shores, and a few swim in tropic latitudes where they also colonize the cold, deep waters. Except for a few varieties that wander into brackish waters, these flattened animals are strictly marine. When not cruising just off the bottom with graceful sweeping strokes of their immense pectoral fins or "wings," rajids rest upon the substrate, or may frequently bury themselves in gravel, sand or silt to conceal themselves from both predator and prey.

Seasonal spawning periods bring mature male and female rajids together for breeding behaviour that culminates in actual copulation. Soon after she is fertilized, special glands within the pregnant female secrete a chitonous, horny envelope, called an egg case, around one or more of the large yolky fertile eggs. Several months pass before the mother skate deposits the distinctive, species-specific egg cases onto the sea floor; during upcoming weeks, final development of each baby skate occurs as it lives off an ever-dwindling yolk supply. Ultimately the young skates, miniature replicas of the adult, wriggle free of the weakened egg cases and begin to explore their new environment.

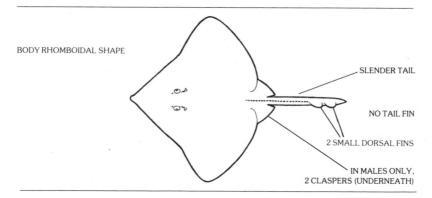

BODY RHOMBOIDAL SHAPE

SLENDER TAIL

NO TAIL FIN

2 SMALL DORSAL FINS

IN MALES ONLY,
2 CLASPERS (UNDERNEATH)

Fossil-bearing rocks containing hardened remains of prehistoric skates have been unearthed, and paleontologists believe them to have existed during upper Cretaceous times, at least 63 million years ago. Today other dorso-ventrally flattened cartilaginous fishes such as rays, stingrays and guitarfishes ally most closely with the rajids. Although plentiful and excellent to eat, skates are still seldom pursued by the oceans' harvesters, and end up in the market place in only limited, incidental quantities. This underutilization will no doubt change as demands upon the world's protein continue to increase.

Although eight known species of skate inhabit the Pacific Northwest, five of these live in deep water where most people do not readily encounter them: the deepsea skate, the roughscale skate, the California skate, the starry skate—a fish similar to the species on page 15 but possessing several spines around each eye socket—and the broad skate.

7 Longnose Skate

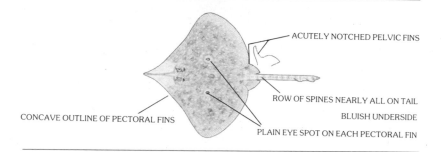

ACUTELY NOTCHED PELVIC FINS

ROW OF SPINES NEARLY ALL ON TAIL

BLUISH UNDERSIDE

PLAIN EYE SPOT ON EACH PECTORAL FIN

CONCAVE OUTLINE OF PECTORAL FINS

SPECIES: *Raja rhina* —from the Latin *raja*, meaning "skate;" and the Greek *rhina*, "nose."

ALTERNATE NAMES: long-nose skate.

MAXIMUM RECORDED SIZE: 137 cm (54 inches).

DISTRIBUTION: Point Loma, southern California, to southeastern Alaska.

While not often seeking the longnose skate, the baitfishing angler occasionally hooks it when fishing over muddy or silty substrates. Pieces of herring, anchovy or shiner, as well as clams, shrimps and marine worms are optimal baits for this skate. Readily attracted to the erratic flashing of silver jigs, the longnose skate may also unintentionally impale or snag itself, but expect no dramatic fight from this lethargic species.

When cruising along sandy, silty bottoms at depths greater than 20 m (66 feet), you must look closely for the longnose skate because it habitually rests partially or completely buried, its sandy coloured body usually difficult to see. If a specimen does swim by, notice that the graceful, sweeping strokes of its pectoral fins provide propulsion while its tail drags behind. Watch also for the longnose skate's egg cases—each 8 to 12 cm in length—that may lie on the sea floor.

When fished over smooth level bottoms, set lines or trawls may capture longnose skate in large numbers at depths to 680 m (2,245 feet). If market prices warrant, retain the "wings" or pectorals and sell them fresh: all too often this excellent seafood is discarded at sea.

The longnose skate lives at depths where the beachcomber or dockside observer cannot see it.

For some truly fine eating, remove and skin the large pectoral fins or "wings" from the longnose skate, then panfry them in butter—also try panfried skate livers. Dubious seafood restaurateurs use cookie cutters to cut skate wings into round pieces and pass them off as scallop.

8 Big Skate

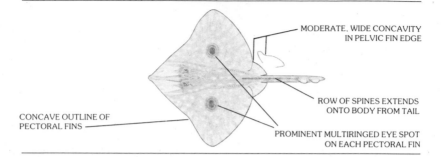

MODERATE, WIDE CONCAVITY IN PELVIC FIN EDGE

ROW OF SPINES EXTENDS ONTO BODY FROM TAIL

CONCAVE OUTLINE OF PECTORAL FINS

PROMINENT MULTIRINGED EYE SPOT ON EACH PECTORAL FIN

SPECIES: *Raja binoculata*—from the Latin *raja*, *bi*, and *oculata*; "skate," "two," and "eyed."

ALTERNATE NAMES: Pacific great skate, Pacific barndoor skate, barndoor skate*, barn-door skate*.

MAXIMUM RECORDED SIZE: 2.4 m (8 feet) and 91 kg (200 pounds).

DISTRIBUTION: Bahia San Quintin, northern Baja California, Mexico, to the Bering Sea coast of Alaska.

 Because it may inhabit depths to within a few metres of the surface, the big skate is the one most often caught. An often surprising and novel catch, this species may seize baits such as clams, shrimps, marine worms and fishes offered over sandy, nearly level bottoms. Irregular flashing of shiny jigs or lures may attract this skate and it becomes foul-hooked more often than conventionally caught. No matter how it is taken, it usually puts up only a sluggish resistance.

 Although the broad, big skate inhabits shallow, easily explored depths, it is well camouflaged against gray sandy or muddy bottoms. So look closely! Attracted by the beast's large size, spearfishermen occasionally shoot this skate, but may perhaps first have to find it partially or completely buried.

 An incidental catch during many commercial harvesting activities, the big skate frequently finds its way to the market place, as "skate wings," where it appears as a very tasty, inexpensive buy.

 A big skate occasionally glides over sandy or silty bottoms within view of the surface-bound naturalist; more often though one finds the large, empty mermaid's purses or egg cases washed ashore.

 For a real taste treat boil the skinned pectoral fins or "wings" of the big skate in water with a bit of vinegar added or in a court-bouillon; its boneless flesh has a flavour not unlike scallop.

*—incorrect

9 Sandpaper Skate

NO SPINES AROUND EYES

CONTINUOUS ROW OF SPINES FROM TAIL EXTENDING ALMOST TO EYE REGION

CONVEX OUTLINE OF PECTORAL FINS

ONE OR TWO SPINES ON SHOULDER AREA

SPECIES: ***Bathyraja kincaidi***--from the Latin *bath* and *raja*, meaning "deep" and "skate;" and honouring Professor Trevor Kincaid from the University of Washington.

ALTERNATE NAMES: black skate.

MAXIMUM RECORDED SIZE: 84 cm (33 inches).

DISTRIBUTION: Santa Catalina Island, southern California, to Unalaska Island, in the Aleutian chain, Alaska.

Rarely does the baitfisherman encounter the sandpaper skate because it lives in deep haunts of 60 to 1,500 m (200 to 5,000 feet) and upon muddy or sandy bottoms. Clams, shrimps, or small fishes would undoubtedly make good bait for this fish. Like other skates, this species puts up little effort to escape when hooked.

The sandpaper skate inhabits locales deeper than those to which sport divers would descend: near the 60-m (200-foot) mark and upon a silty, level bottom. Even then, sighting the sandpaper skate would prove difficult because the sandy-coloured beast frequently buries itself, making its form indistinct from its surroundings.

When pursuing valuable groundfish and during their many harvesting activities, trawlers and shrimpers frequently capture the sandpaper skate. Seldom do they market this small skate; instead they usually discard it at sea along with many other unsaleable animals. Small quantities of sandpaper skate occasionally have been rendered to produce mink feed or fertilizer.

Don't expect to see the sandpaper skate because this deepwater denizen never invades water shallow enough for observation from the surface.

A large sandpaper skate is definitely worth preparing, but first remove and retain the broad, wing-like pectoral fins while discarding the remaining carcass. Next, skin the fins. The flesh is flavourful and has a crab-like texture. Try skate au gratin, or make fritters from the carefully hoarded livers. Retain any leftovers too, they make an excellent casserole base.

THE CHIMAERAS
(Family: Chimaeridae)

Ratfishes, chimaeras and ghost sharks make up a small family of grotesque fishes, scientifically termed Chimaeridae, of which approximately 25 present-day species have been described. The various chimaerids live only in the world's oceans and seas; indeed, most flourish in the Pacific and Atlantic Oceans. Only one of the known species, however, resides in the Pacific Northwest.

Although a few chimaerids live in shallow coastal habitats, most inhabit deep and dark offshore haunts. These species seem to become more grotesque as the depth increases: long-nosed, ghostly forms abound at abyssal reaches. Slow, graceful swimmers, all reside near the sea floor, including the largest species—a North Atlantic variety that may attain 152 cm (60 inches) in length.

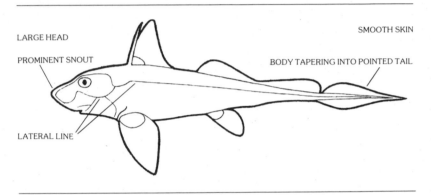

The name chimaera suggests the monsters of Greek mythology, indeed, these fishes are as grotesque as anything unreal or imaginary. They have long rat-like tails and bulbous heads with round or pointed snouts but have almost rabbit-like mouths. Their huge eyes often shine emerald green from a scaleless "metallic" body which shimmers silver and bronze, a form which looks as though its tail end does not match its front end. And they swim as though they are slow-flying birds.

Collected fossils indicate that in early times chimaeras flourished in great diversity and number. Paleontologists trace the chimaerid ancestors back to the upper Devonian period, back 345 million years, but now only the distantly related sharks and rays closely approximate them. Historically chimaerids were more useful to mankind than they are today: before the discovery of new synthetics man processed chimaerid livers to make fine oils for lubricating machinery.

10 Ratfish

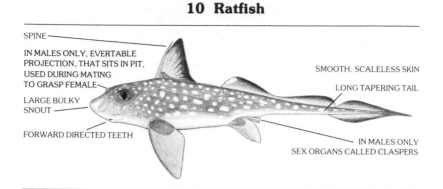

SPINE

IN MALES ONLY, EVERTABLE
PROJECTION, THAT SITS IN PIT,
USED DURING MATING
TO GRASP FEMALE

LARGE BULKY
SNOUT

FORWARD DIRECTED TEETH

SMOOTH, SCALELESS SKIN

LONG TAPERING TAIL

IN MALES ONLY
SEX ORGANS CALLED CLASPERS

SPECIES: Hydrolagus colliei—from the Greek *hydor* and *lagus*, meaning "water" and "hare;" and honouring Alexander Collie, ship's surgeon and early amateur naturalist with Captain Beechey's *Blossom*.

ALTERNATE NAMES: spotted ratfish, rat-fish, chimaera, rabbitfish, spookfish, elephant-fish, goatfish, water hare.

MAXIMUM RECORDED SIZE: 97 cm (38 inches).

DISTRIBUTION: Isla Cedros, central Baja California, Mexico, to Cape Spencer, southeastern Alaska; also an isolated population at Isla Tiburon, in the upper Gulf of California, Mexico.

The mostly unwanted ratfish often bites gently at shrimps, crabs, snails, piling worms, or small fishes offered by bottomfishermen. More frequently, though, this grotesque creature accidentally impales one of its large, winglike pectoral fins on the hook. Because of its long slender tail, the ratfish is a poor fighter and is therefore not popular with sportsmen. Besides, most anglers consider it ugly because its soft body sags when lifted free of the water.

When viewed underwater, particularly at night, the spectacular ratfish immediately attracts the diver's attention. A shimmering silver and bronze form glides slowly over the bottom, propelled by bird-like sweeps of its pectoral fins. At night, the young and the adults swarm over sandy bottoms in shallow water. Empty ratfish egg cases—dark and spindle-shaped—commonly rest upon the sea floor.

Considered a time-wasting nuisance by modern commercial fishermen, the abundant ratfish—currently so economically valueless—is a very common catch, and particularly plagues deep dragging trawlers and deep setting longliners.

Look for ratfishes where a shore or a wharf light shines into the water because they seem attracted to light and may rise from deep water toward it. Shorebound naturalists rarely see this species, although it is very occasionally trapped in tidepools.

Its texture is like crab, but ratfish flesh leaves an unpalatable aftertaste.

THE STURGEONS
(Family: Acipenseridae)

A simple glance at the heavily armoured, shark-like form of a sturgeon indicates the primitive nature of the Acipenseridae. Only 25 known species still exist. Living sturgeons are restricted to the cool waters of the northern hemisphere, with habitats in Eurasia, particularly, containing most of the acipenserids. For at least some portion of their life cycles, all acipenserids live in fresh water, and most species spend their entire lives in large silty-bottomed rivers or lakes. A few types, though, may migrate to sea as adult fish only to return ultimately to fresh water for spawning—they are called anadromous species.

The slow but steadily swimming sturgeons live primarily in very murky water where sight assists little in locating bottom-dwelling prey. Consequently the tiny-eyed acipenserids compensate for poor visibility with external taste buds located on the long, whisker-like barbels that precede their mouths. By dragging these sensitive barbels along the bottom the sturgeon locates small animals, such as shrimp, worms and clams before sucking them up into their protrusible, tube-like mouths. When searching for food, sturgeons propel themselves with slow, powerful sweeps of their primitive heterocercal tails—tails in which the backbones extend into the upper lobes giving them an unbalanced appearance.

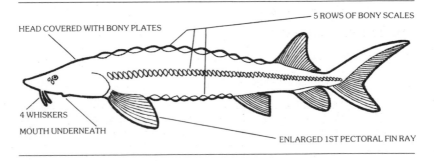

HEAD COVERED WITH BONY PLATES

5 ROWS OF BONY SCALES

4 WHISKERS

MOUTH UNDERNEATH

ENLARGED 1ST PECTORAL FIN RAY

Early literature notes some truly gargantuan sturgeons and nearly all species may grow to 152 cm (5 feet). Indeed, the world's largest freshwater fish is an acipenserid—the beluga of northern Asia which may attain a length of 8.2 m (28 feet) and a weight of 1,360 kg (3,000 pounds)! For years mankind has utilized the massive sturgeons for their renowned caviar, which is actually processed eggs stripped from mature females. A liquid once used by early breweries to clarify beer, isinglass, was prepared from sturgeon swim bladders.

Although two sturgeon species live in the Pacific Northwest, and primarily occupy the same river systems, only one is commonly observed. The seldom-seen green sturgeon with its distinctive olive coloration, its white lateral stripes and its 23 to 30 lateral plates along each side, receives no further attention in this volume.

18

11 White Sturgeon

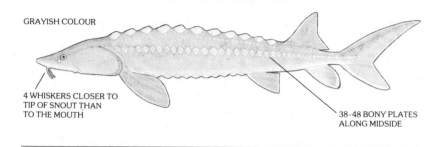

GRAYISH COLOUR

4 WHISKERS CLOSER TO
TIP OF SNOUT THAN
TO THE MOUTH

38-48 BONY PLATES
ALONG MIDSIDE

SPECIES: *Acipenser transmontanus*—from the Latin *acipenser*, *trans* and *montanus*; "sturgeon," "beyond" and "mountain."

ALTERNATE NAMES: Pacific sturgeon, Oregon sturgeon, Columbia sturgeon, Sacramento sturgeon.

MAXIMUM RECORDED SIZE: 6.1 m (20 feet) and 816 kg (1,800 pounds).

DISTRIBUTION: northern Baja California, Mexico, to Cook Inlet, Gulf of Alaska.

Baitfishing along the banks of large rivers, such as the Fraser and Columbia, accounts for the bulk of the white sturgeon sport catch. In such locales, try bottomfishing with worms, or even lampreys easily dug from adjacent small creeks, but obey all special local regulations before taking sturgeon. Marine anglers who troll or stillfish near estuaries occasionally capture these strong creatures which put up a tough, long battle, sometimes with aerial acrobatics. Special heavy tackle is needed to catch large specimens.

Few divers venture into turbid, current-swept rivers or estuaries where the white sturgeon is plentiful, but where viewing conditions are extremely poor.

A long, steadily continuing fishery for the white sturgeon survives within the lower regions of major river systems, where the take is usually incidental for gillnetters pursuing salmon. Marine trawlers and seiners, too, periodically happen upon commercial quantities of white sturgeon. Fishermen market the excellent flesh, but also receive good prices for its roe which, when processed, sells as caviar.

Only when a disturbed white sturgeon leaps clear of the very silty waters of rivers or their estuaries is the shorebound naturalist likely to observe it.

Fine dining awaits anyone who samples the white sturgeon. Its flaky, white flesh contains virtually no bones and cuts beautifully into portion-sized steaks. Marinate them in oil, lemon juice, paprika and spices. Try curried sturgeon. Or why not stuff it with truffles, marinate it in brandy, and poach in champagne?

19

THE HERRINGS
(Family: Clupeidae)

The very large herring family, scientifically termed Clupeidae, contains the schooling shads, alewives, menhaden, sardines and herrings. Except for the frigid marine habitats of the polar regions, the seas of today's world contain the more than one hundred and ninety known clupeids, with tropical regions, such as the vast expanse of the Indo-Pacific, harbouring most of them. Although certain varieties tolerate brackish water at some point during their life cycles, strictly freshwater herrings are very few. These silvery fishes, most of which attain lengths less than 46 cm (18 inches), school in open shallow regions or in midwater over deeper haunts and not uncommonly gather in immense shoals that cover many square kilometres of ocean.

Today's clupeid fishes apparently differ little from their Cretaceous forbears which lived 110 million years ago. The anchovies are now most closely related to the herrings; species in both groups are streamlined, silvery, schooling fishes that lack lateral line canals and small fleshy adipose fins. Economies of many nations, both directly and indirectly, depend very significantly on the various existing herring fisheries of the world. Seining and gillnetting annually account for millions of kilograms of clupeid protein from fisheries that not only feed mankind directly, but provide bait for other fishing enterprises as well as ingredients for fertilizers. Because other fish feed upon herrings, their indirect worth defies evaluation. Their double importance, therefore, makes an awesome responsibility for fisheries management personnel.

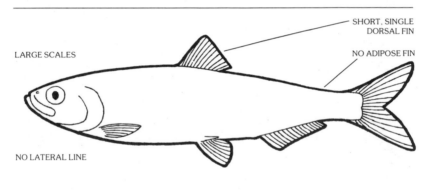

SHORT, SINGLE DORSAL FIN

NO ADIPOSE FIN

LARGE SCALES

NO LATERAL LINE

Aside from the common Pacific herring, two other clupeid fishes live in the Pacific Northwest. The world's largest herring, the American shad, at 76 cm (36 inches), originally introduced here from North America's Atlantic seaboard has successfully colonized this coast and takes lures in some areas while migrating upriver to spawn. Popularly known as the "pilchard," the native Pacific sardine, once itself the object of a large fishery but now mysteriously rare in the Pacific Northwest, resembles the herring but possesses special large scales on the sides of its tail as well as a series of spots down each side.

12 Pacific Herring

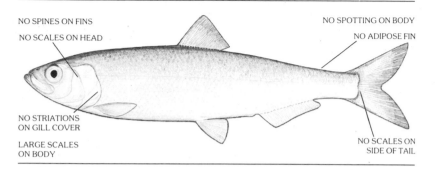

NO SPINES ON FINS

NO SCALES ON HEAD

NO SPOTTING ON BODY

NO ADIPOSE FIN

NO STRIATIONS
ON GILL COVER

LARGE SCALES
ON BODY

NO SCALES ON
SIDE OF TAIL

SPECIES: *Clupea harengus pallasi*—from the Latin *clupea*, meaning "herring;" *harengus*, an ancient word for "schooling;" and honouring famous Russian naturalist Petrus Pallus, with a sub-species name that distinguishes this form from its close relative in the Atlantic.

ALTERNATE NAMES: herring.

MAXIMUM RECORDED SIZE: 46 cm (18 inches).

DISTRIBUTION: northern Baja California, Mexico, to the Bering Sea and northeast to Bathhurst in the Beaufort Sea; along the Asian coast from the Lena River in the Arctic Ocean, through the Bering Sea to Korea and Ibaraki, central Japan.

The foremost baitfish in the Pacific Northwest, the Pacific herring is particularly valuable to salmon anglers who fillet it into "strip" for trolling and casting, use it whole for trolling, or alive for a type of still-fishing called "mooching." Resourceful sportsmen often capture their own Pacific herring by "raking" or jigging their various bare-hooked or sharp-pronged instruments.

Large schools of Pacific herring, particularly juveniles, crowd into shallow bays or along shores which are popular spots for divers. Look up into the greenish open water to see a school as a silvery mass moving as if it were one animal until your startling approach temporarily scatters the unit.

Particularly just after the Second World War, an immense seine fishery for Pacific herring developed in the Pacific Northwest. While the bulk of this catch was formerly reduced to meal, other more profitable products such as kippers, bloaters, roll mops, canned and dry-salted herring still remain popular. Most recently the bonanza seine and gillnet roe fishery ominously threatens many other local fishing operations because so many commercially important fish feed very heavily upon the Pacific herring.

The small silvery fish "flipping," or leaping clear of the water and turning on their sides under the surface, creating a silvery flash, are nearly always Pacific herring.

A multi-use seafood, the Pacific herring is delicious fresh. Try poaching herring roe; add a bit of vinegar or lemon juice.

THE ANCHOVIES
(Family: Engraulidae)

Approximately one hundred and twenty-five known living species form a large family of fishes called the Engraulidae or anchovies. These streamlined, silvery fish actively gather in surface and midwater habitats, often forming truly immense schools to feed, breed and migrate en masse. Most of these small—usually less than 12-cm (5-inch)—fishes shoal in warm marine waters. Although some engraulids tolerate brackish waters, few live in strictly freshwater realms. Indo-Pacific and southern Atlantic regions particularly, possess a wealth of anchovy species, while only one, a veritable giant at 24.8 cm (9.8 inches), lives in the Pacific Northwest.

An anchovy feeds by grazing plankton—tiny floating plants and animals—from oceanic pastures. After gulping a large amount of plankton-containing water, an anchovy, by means of long, thin, close-packed gill rakers, strains the food from the water as it passes through the gills and out the gill openings or opercula. Mucous then forms on these gill rakers and envelops the food; this nutritional bolus then sloughs off and slides down the throat to the fish's stomach.

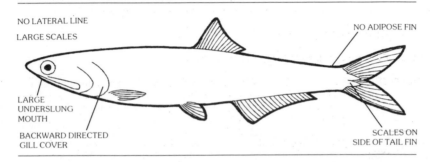

NO LATERAL LINE

LARGE SCALES

NO ADIPOSE FIN

LARGE UNDERSLUNG MOUTH

BACKWARD DIRECTED GILL COVER

SCALES ON SIDE OF TAIL FIN

By aging the most ancient engraulid fossils yet unearthed, paleontologists believe ancestral anchovies existed as long as 60 million years ago, in Tertiary times. Inspection of the world's present-day fish fauna reveals the anchovies' closest living kin to be the herrings: member species of both families are silvery and streamlined with large, easily removed scales and fleshy flaps just preceding their tails, but lack lateral lines. In many regions, harvesting very significant anchovy populations not only directly supplies protein for human consumption, but also provides valuable bait for other important fisheries. Even more noteworthy, though, engraulids feed countless other creatures, and therefore prove vital to the ecology of tropical seas in particular.

13 Northern Anchovy

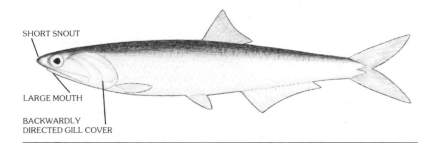

SHORT SNOUT

LARGE MOUTH

BACKWARDLY
DIRECTED GILL COVER

SPECIES: ***Engraulis mordax***—from the Greek *engraulis*, meaning "anchovy;" and the Latin *mordax*, "biting."

ALTERNATE NAMES: plain anchovy, California anchovy, anchovy.

MAXIMUM RECORDED SIZE: 24.8 cm (9.8 inches).

DISTRIBUTION: La Paz, in the Gulf of California, around the tip of Baja California, Mexico and north to the Queen Charlotte Islands, northern British Columbia.

 Anglers, particularly those along the Californian and Mexican coasts, extensively use the northern anchovy as bait, often dropping large quantities from party boats to "chum" gamefish and to encourage them to remain nearby. Hooked living or dead anchovy soon follow as offerings to the now frenzied sportfish. In the Pacific Northwest increasingly more salmon fishermen are using this anchovy as bait.

 Particularly in southern regions, huge schools of northern anchovy often swarm in popular dive sites, but individuals are difficult to approach because they quickly dart away at the slightest disturbance. Cameras mounted on submersibles have filmed this anchovy at depths as great as 310 m (1,023 feet).

 Although an immense northern anchovy fishery in California, harvested by surface-fishing purse seiners, fuels an important reduction industry, only a small part of this take is selected for human consumption, either canned or fresh. Lampara nets capture a sizable bait-take of four year old or younger northern anchovy.

 Northern anchovies occur sporadically in large schools around wharves. Watch a specimen as it feeds and see it open its large mouth to seize a planktonic morsel. During feeding, the fish's lower jaw drops and its head seems almost to fall apart—a spectacular sight when an entire school is foraging in unison.

 Canned northern anchovies have a fine, delicate flavour, while fresh cooked specimens compare favourably with delicious juvenile Pacific herring. Fry in olive oil.

23

THE SMELTS
(Family: Osmeridae)

A small family of approximately twelve known living species, the Osmeridae, meaning "odorous family" in Greek, contains the smelts, the capelin and the eulachon. Temperate and boreal waters of the North Pacific Ocean harbour most smelts, with seven of the known species living in the Pacific Northwest; the remaining forms thrive in the Arctic and Atlantic Oceans. While most live out their entire lives in shallow coastal marine habitats, a few more adaptable osmerids migrate into rivers or streams to breed and these species are called anadromous. After the fertilized eggs of these anadromous species develop and hatch, the tiny, weak-swimming transparent larvae are swept downstream to sea where they begin to feed and grow rapidly. Osmerids, most of which attain less than 30 cm (12 inches) in length, swim in surface or midwater layers, and usually form schools.

After collecting and studying the impressions or remains of readily recognizable, but extinct smelt, paleontologists believe that these creatures abounded in lower Eocene seas some 40 million years ago. Within today's fish fauna, the salmon and trout relate most closely to the osmerids; members of both families possess adipose fins and several noticeable skeletal similarities.

LATERAL LINE

ADIPOSE FIN

NO LARGE POINTED
SCALE ABOVE EACH
PELVIC FIN

FORKED TAIL FIN

Modest but locally significant commercial fisheries harvest some of the larger varieties of smelt and usually these economic activities coincide with the spawning seasons, when these fish congregate along shallow accessible beaches or in rivers and estuaries. As significant prey for many larger animals, such as marine mammals, sea birds and economically valuable fishes, osmerids are of immense indirect importance to humanity.

While present in the Pacific Northwest, the whitebait smelt, the rainbow smelt, and the night smelt—a fish similar to the species on page 28 but with shorter pectoral fins—are not encountered by anglers, divers, or naturalists.

24

14 Eulachon

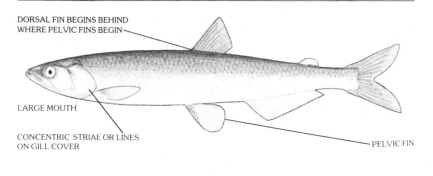

DORSAL FIN BEGINS BEHIND
WHERE PELVIC FINS BEGIN

LARGE MOUTH

CONCENTRIC STRIAE OR LINES
ON GILL COVER

PELVIC FIN

SPECIES: *Thaleichthys pacificus*—from the Greek *thaleia* and *ichthys*, meaning "rich" and "fish;" and emphasizing the Pacific Ocean.

ALTERNATE NAMES: Columbia river smelt, candlefish, oilfish, fathom fish, salvation fish, small fish, yshuh, swaive, chucka and numerous spelling or phonetic variants of eulachon: oolakon, oolachon, oolichan, oulachon, oulachan, oulacon, ulchen, uthlecan.

MAXIMUM RECORDED SIZE: 23 cm (10 inches).

DISTRIBUTION: Russian River, central California, to Bristol Bay on the Bering Sea coast of Alaska and the Pribilof Islands.

Although anglers do not seek or catch the small-mouthed eulachon, it offers excellent but as yet unexploited potential as bait.

Only when the anadromous eulachon makes its spring spawning migration to large silty rivers does this oily fish swim in shallow diveable waters; however, turbidity near large river mouths would obscure this schooling fish from view.

In the marine environment, the eulachon blunders into commercial trawl nets only occasionally, but once this smelt enters its spawning rivers, it is briefly but intensely gillnetted and sold in local markets.

Look along the estuaries and banks of large rivers, such as the Fraser or Columbia, after the eulachon's spring spawning rituals to find many dying spawned-out specimens as they drift downstream. Many hungry and knowledgeable people wander along the banks to gather these specimens during a two or three-week period.

Because of its high oil content, native Indians along large rivers have trapped, raked or netted the highly-prized eulachon for centuries. These fish were vital to the aboriginal economy not only because they represented food and fat but also because they were items of barter along the "grease trail" to the Interior. Today many Pacific Northwesterners actively seek the popular eulachon for frying in butter.

15 Surf Smelt

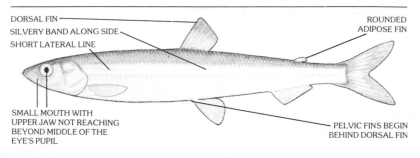

DORSAL FIN

SILVERY BAND ALONG SIDE

SHORT LATERAL LINE

ROUNDED ADIPOSE FIN

SMALL MOUTH WITH UPPER JAW NOT REACHING BEYOND MIDDLE OF THE EYE'S PUPIL

PELVIC FINS BEGIN BEHIND DORSAL FIN

SPECIES: *Hypomesus pretiosus pretiosus*—from the Greek *hypo* and *mesus*, meaning "below" and "middle;" and the Latin *pretiosus*, "precious"—the second *pretiosus* distinguishes this sub-species from a close relative living along Asian shores.

ALTERNATE NAMES: silver smelt, smelt.

MAXIMUM RECORDED SIZE: 30.5 cm (12 inches).

DISTRIBUTION: Long Beach Harbour, southern California, to Chignik Lagoon, on the Alaskan Peninsula.

In some localities, a certain kind of angler catches surf smelt from wharves or piers with a "smelt jig," a line from which he dangles a series of shiny, bare hooks. Laconner, Washington has sponsored an annual smelt jigging derby on the first Saturday of February, an action-packed family festival. Bait fishermen can find the surf smelt to be an excellent lure.

A really observant diver seldom sees the surf smelt because it frequents the murky waters of shallow beaches. An improbable sighting would only be minimal in any case and not suitable for positive identification.

Commercial fishermen take the abundant surf smelt with gill nets or seine nets set from sandy beaches and they market it fresh. Optimal harvesting occurs during summer in British Columbia and Washington. The catch is modest, seasonal, and localized, as are sales.

In summer, when the spawning surf smelt swarms to sandy beaches, enthusiastic amateur "smelters" do the same, gathering the tasty fish with gillnets or dipnets, rakes, buckets, and even hands. Try twilight hours, just before a high tide, after first consulting local regulations. Native Indians, such as the Quillehutes on the outer coast of the Olympic Peninsula, welcome the seasonal appearance of the surf smelt, and much tribal lore focuses on these schooling fish which are so easy to gather.

Most aficionados fry flour-covered surf smelt whole in butter. Gut them or not. Beach parties, especially those with open fire cookery, truly enhance surf smelt collecting trips. The Quillehutes dried these white-fleshed fish on cedar bark.

16 Capelin

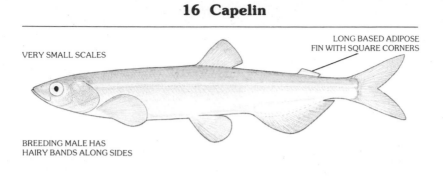

VERY SMALL SCALES

LONG BASED ADIPOSE
FIN WITH SQUARE CORNERS

BREEDING MALE HAS
HAIRY BANDS ALONG SIDES

SPECIES: *Mallotus villosus*—from the Greek *mallotus*, and the Latin *villosus*, both meaning "hairy."

ALTERNATE NAMES: Pacific capelin, lodde.

MAXIMUM RECORDED SIZE: 21.8 cm (8.6 inches), in the Pacific Ocean.

DISTRIBUTION: Juan de Fuca Strait, southern British Columbia, to Arctic Alaska and along the Asian Pacific coast from Kamchatka, USSR, to Hokkaido, northern Japan and the Tumen River, Korea. In the Atlantic Ocean from Labrador to Greenland, Iceland, Norway and Novaya Zemlya, USSR.

A chief dietary item for many fishes, the schooling, silvery capelin is undoubtedly an excellent bait but, except for a few weeks in autumn, this small smelt lives in deep waters inaccessible to the bait-seeking angler.

An opportunity for Pacific Northwest divers to view the seldom-seen capelin arrives during September and October when it swarms inshore to spawn along certain shallow, gravelly beaches: dive nocturnally, at high tide, when the moon is full.

In spite of its accessibility during shallow water autumn spawning appearances, the capelin is yet to be commercially exploited in the Pacific Northwest. Throughout the rest of the year trawlers catch small incidental quantities.

By the light of a full moon, during September and October, dedicated enthusiasts eagerly gather along selected gravelly beaches for the spawning capelin. When these swarming smelt actually wriggle out of the water to spawn among the pebbles and sand, collectors scoop them up with buckets, rakes, screening, nets, or even by hand. Each 2 or 3 year old female capelin may deposit up to 6,000 tiny adhesive eggs which stick to the gravel before being buried by subsequent waves. The young develop and are ready to hatch at the next equally high tidal sequence in about two weeks. Beautiful biological timing!

Capelin lovers fry their catches whole, often right on the beach over a bonfire.

17 Longfin Smelt

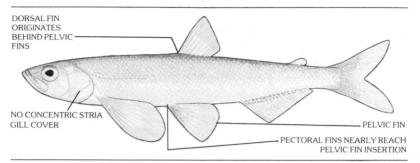

DORSAL FIN ORIGINATES BEHIND PELVIC FINS

NO CONCENTRIC STRIA GILL COVER

PELVIC FIN

PECTORAL FINS NEARLY REACH PELVIC FIN INSERTION

SPECIES: *Spirinchus thaleichthys*—from the Latin *spirinchus*, meaning "smelt;" and the Greek *thaleia* and *ichthys*, "rich" and "fish."

ALTERNATE NAMES: long-finned smelt, Puget Sound smelt, Pacific smelt.

MAXIMUM RECORDED SIZE: 15.2 cm (6 inches).

DISTRIBUTION: San Francisco Bay, central California to Prince William Sound, Gulf of Alaska. Isolated populations also flourish in fresh water localities such as Harrison Lake, British Columbia and Lake Washington, Washington.

Occasionally while jigging with special rigs consisting of several small, shiny tandemly-arranged hooks, the angler might take a longfin smelt. A forage species for many large and desirable sportfish, this delicate-looking smelt offers great promise as bait and a potential bait fishery seems obvious.

In autumn the longfin smelt moves into shallow beach locales, but even then observant divers seldom see this active schooling fish, usually because of turbid conditions.

Commercial gillnetters who harvest other smelt rarely take the small longfin smelt because it easily slips through the larger mesh size. Shrimp trawlers, though, with their finer-meshed nets dragging the bottom down to 135 m (450 feet), sporadically capture this plankton-consuming smelt as an incidental or unwanted catch. An important food source for many prominent commercial fishes, this smelt matures and spawns at two years of age, with very few surviving to breed again the following season.

Blessed with clear water, the beachcomber or dockside observer might catch a fleeting glimpse of longfin smelt—but not good enough for a positive identification, especially from above water.

Soak in milk, coat with flour and deep fry. Eat them right away or marinate twelve hours in vinegar with onion, thyme and bay leaf. Try smelt shishkebab. Smelt require salt.

THE SANDLANCES
(Family: Ammodytidae)

Scientifically termed Ammodytidae, the sandlances comprise a small, very distinct family of fishes containing approximately twelve recognizable present-day species. When not actively schooling in large aggregations, usually within a few metres of the surface, these long, thin creatures bury themselves tail first in the fine sand of nearby, shallow shores, a behaviour obviously prompting the popular name sandlance. Members of this strictly marine family are distributed widely and fairly evenly throughout the world's oceans and seas, with representative species in tropical, temperate and even boreal latitudes.

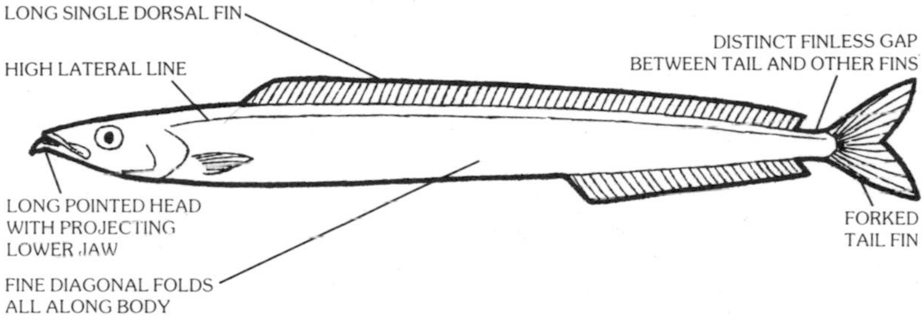

LONG SINGLE DORSAL FIN

HIGH LATERAL LINE

DISTINCT FINLESS GAP
BETWEEN TAIL AND OTHER FINS

LONG POINTED HEAD
WITH PROJECTING
LOWER JAW

FORKED
TAIL FIN

FINE DIAGONAL FOLDS
ALL ALONG BODY

Judging by what has been uncovered, prehistoric sandlances left only a little fossilized material behind, but these scanty remains indicate that ancient ammodytids lived in shallow marine seas at least 25 million years ago, during the Oligocene epoch. After years of studying sandlance characteristics and comparing them with those of other present-day fishes, scientists believe the Ammodytidae is a very distinct group, not closely related to any other. Though serious exploitation of these tiny creatures—most grow to less than 30 cm (12 inches)—does not occur, they indirectly affect other fisheries because of their immense value as forage for many commercially valuable species.

18 Pacific Sand Lance

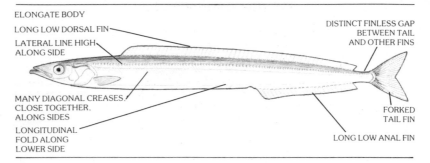

ELONGATE BODY

LONG LOW DORSAL FIN

LATERAL LINE HIGH ALONG SIDE

MANY DIAGONAL CREASES CLOSE TOGETHER. ALONG SIDES

LONGITUDINAL FOLD ALONG LOWER SIDE

DISTINCT FINLESS GAP BETWEEN TAIL AND OTHER FINS

FORKED TAIL FIN

LONG LOW ANAL FIN

SPECIES: *Ammodytes hexapterus* —from the Greek *ammos*, *dyo*, *hex*, and *pteros*, meaning "sand," "burrow," "six" and "wing."

ALTERNATE NAMES: Pacific sandlance, arctic sandlance, sand-lance, sandlaunce, needlefish.

MAXIMUM RECORDED SIZE: 27 cm (10.5 inches).

DISTRIBUTION: Balboa Island, southern California, to the Bering Sea and east along the Arctic Ocean shore to Clearwater River, Hudson's Bay. Also south along the Asian shore to the Sea of Japan.

A very important prey for many large inshore fishes, the small Pacific sand lance ranks potentially as an excellent bait, especially when used live.

Particularly in summer, divers snorkeling along the surface often encounter vast schools of Pacific sand lance near shore. These schooling fish scatter before the diver and make close viewing difficult.

The small, pencil-thin, unmarketable Pacific sand lance easily swims through any commercial netting, and therefore avoids incidental capture.

In spring a Pacific sand lance, 8 years old or more, leaves the bottom to breed at the surface and may form, with other younger specimens, large densely packed schools near wharves and pilings. At night, when attracted by a beam of light, these swarming creatures may be easily scooped up with a long-handled dipnet. Although shorter deeper-bodied juvenile herring commonly infiltrate such schools, the slender, elongate Pacific sand lance may be easily distinguished by its "eel-like" swimming motion. While strolling along a sandy shore at low tide, barefooted beachcombers may be startled to find specimens squirming out of the sediment, and perhaps even wriggling between their toes. Digging may expose many more.

For a delicious, crunchy snack or hors d'oeuvre, after gathering a sufficient quantity of fresh Pacific sand lance, coat them with flour and fry them whole in butter.

THE SALMON AND TROUT
(Family: Salmonidae)

Not only does the family Salmonidae include the salmon and trout, but also the chars. Active fishes that often school, the more than sixty known living species of salmonids may occur in salt or fresh water, with some species spending at least part of their lives in each environment. These adaptable species that, as adults, migrate from the sea back to rivers and streams to spawn are anadromous, and include eight Pacific Northwest species. Although many salmon and trout live only in fresh water throughout their lives, no completely marine species exist. Originally, the boreal and temperate waters of the northern hemisphere contained all of the world's salmon and trout, but man has extensively transplanted many of these popular fishes into nearly all waters of the globe. For example, the hardy rainbow trout, originally native only to North America, flourishes today in virtually every fresh water corner of the world. Much to man's delight, many salmonids attain impressive sizes, with the Pacific Northwest's chinook salmon topping the list at 147 cm (58 inches).

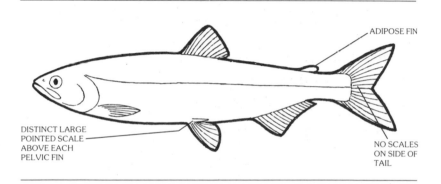

ADIPOSE FIN

DISTINCT LARGE
POINTED SCALE
ABOVE EACH
PELVIC FIN

NO SCALES
ON SIDE OF
TAIL

Prehistoric ancestors of present-day salmonids lived at least 40 million years ago during the lower Eocene epoch. The various smelts, which also possess adipose fins and have certain skeletal similarities with the salmonids, are among the closest living relatives of the salmon and trout. World renowned as gamefish and food sources, salmonids represent a most significant economic factor in many countries; fish farming and ocean ranching offer great promise for the future.Commercial and recreational interests lobby so intensely that governments annually spend millions of dollars to maintain or increase their present salmon and trout stocks, Canada's ambitious Salmonid Enhancement Program being the latest, and perhaps most sophisticated of these efforts.

Two non-resident salmonid fishes have been introduced into the Pacific Northwest, but neither has apparently established marine migrating populations that are viable. The brown trout, a golden brown fish with dark spots surrounded by light halos, exists in some coastal British Columbia stream headwaters, but the Atlantic salmon seemingly never did establish itself to any extent in the Pacific Northwest.

19 Pink Salmon

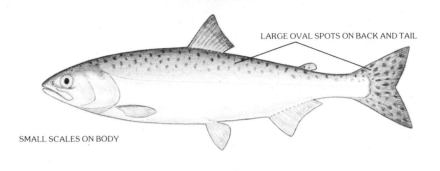

LARGE OVAL SPOTS ON BACK AND TAIL

SMALL SCALES ON BODY

SPECIES: *Oncorhynchus gorbuscha*—from the Greek *onkos* and *rhynchos*, meaning "hook" and "nose;" and the Russian name, *gorbuscha*, for this fish in Alaska.

ALTERNATE NAMES: humpback salmon, pink, humpie.

MAXIMUM RECORDED SIZE: 76 cm (30 inches) and 6.8 kg (15 pounds).

DISTRIBUTION: La Jolla, southern California, to the Bering Sea and the Canadian Arctic to the Mackenzie River; along the Asian coast from the Lena River, Arctic Ocean, through the Bering Sea to the Tumen river, Korea, and Hondo Island, southern Japan. Introductions have been made throughout North America and the Kola Peninsula, northern Europe, but populations have only successfully established themselves in the Great Lakes.

 In some northern locales, the smallish pink salmon is popular with spin casters using "buzz bombs" or "sting sildas" in late summer. Trollers who drag plugs and dodger-spoon rigs also take it.

 The elusive pink salmon adult easily avoids most divers; and an aquanaut who locates a specimen usually only has a fleeting glimpse of it. Along shorelines in May and June, however, tiny oceanbound fry school to feed at the surface where snorklers easily find them.

 While gillnetters and purse seiners take most of the large pink salmon catch for canning, small quantities marketed fresh come via the troller fleet.

 Large schools of tiny, blue-backed pink salmon fry congregate along rocky shores and adjacent to wharves or jetties, but unfortunately the topside observer will find it difficult to distinguish this species from the very similar but parr-marked chum fry.

 Although the pink salmon has historically rated low in comparison with other salmon, it is unquestionably a fine eating fish and when used promptly, worthy of a chef's best efforts; its light pink-coloured flesh bakes, broils, fries or boils very well, suiting virtually any fish recipe. Try a salmon souffle.

20 Chum Salmon

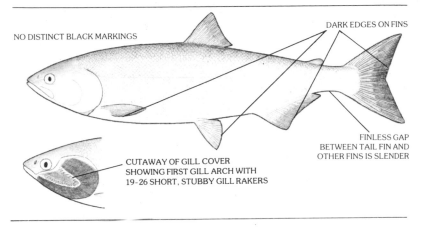

NO DISTINCT BLACK MARKINGS

DARK EDGES ON FINS

FINLESS GAP
BETWEEN TAIL FIN AND
OTHER FINS IS SLENDER

CUTAWAY OF GILL COVER
SHOWING FIRST GILL ARCH WITH
19-26 SHORT, STUBBY GILL RAKERS

SPECIES: *Oncorhynchus keta*—from the Greek *onkos* and *rhynchos*, meaning "hook" and "nose;" and the Russian name, *keta*, for this fish.

ALTERNATE NAMES: dog salmon, fall salmon, keta salmon, calico salmon, chum, qualla.

MAXIMUM RECORDED SIZE: 102 cm (40 inches) and 15 kg (33 pounds).

DISTRIBUTION: San Diego, southern California, to the Bering Sea through to the Mackenzie River, in the Canadian Arctic; along the Asian coasts from the Lena River, Arctic Ocean, through the Bering Sea to the Tumen River, Korea, and Honshu, southern Japan. Transplants to Hudson's and James Bays met with poor results.

The infrequently sought chum salmon occasionally strikes the angler's line; a buzz bomb or a double dodger rig trailing a hook wrapped with yellow yarn particularly fascinates this fish when on its autumn approach to spawning streams. A large specimen can provide a good scrap!

While divers seldom get close enough for even a fleeting glimpse of an adult chum salmon, in spring schools of fry swim at the surface where snorkelers roam.

Marketed canned, smoked, frozen, or more recently, as fresh product, the chum salmon is increasing in popularity. Although it ranks lowest among the Pacific salmon in North America, this species is very important in Asia where the resourceful Japanese can or salt the flesh, and pickle the roe as sujiko.

After leaving their streams of birth in schools, tiny parr-marked chum salmon fry swarm along the sea shore on their way to the open ocean. The pierside naturalist or shorebound stroller, while readily able to see these fish, will have difficulty in distinguishing them from the non-barred pinks.

Of the five Pacific salmon, the chum smokes best because of its high oil content. It's excellent cooked fresh too!

21 Sockeye Salmon

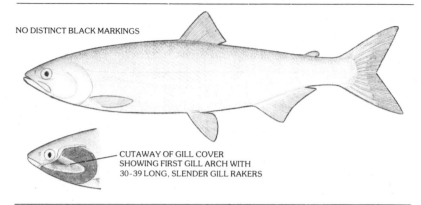

NO DISTINCT BLACK MARKINGS

CUTAWAY OF GILL COVER
SHOWING FIRST GILL ARCH WITH
30-39 LONG, SLENDER GILL RAKERS

SPECIES: *Oncorhynchus nerka*—from the Greek *onkos* and *rhynchos*, meaning "hook" and "nose;" and the Russian popular name for the fish—*nerka*.

ALTERNATE NAMES: red salmon, blueback salmon, sockeye, blue back.

MAXIMUM RECORDED SIZE: 84 cm (33 inches) and 6.8 kg (15 pounds).

DISTRIBUTION: Los Angeles, southern California, to the Bering Sea and north to Bathhurst Inlet on the Canadian Arctic coast; also along the Asian Pacific coast from the Anadyr River to northern Hokkaido, Japan.

Some sockeye salmon take a sport lure, and most of those which do seize squid-like objects or hooks wrapped with red cotton to simulate small planktonic prey. Wrap a hook and try it!

Both adult and juvenile sockeye salmon swim swiftly and so elusively that divers very seldom see them.

Without doubt, the sockeye salmon ranks as the most important commercial fish in the Pacific Northwest—both now and in the past. Domestic seiners harvest along coastal channels and regional gillnetters take their share at river mouths. Although most of the total catch is canned, in recent years trollers tow "egg spoons," "feathered lures," and "hoochies" on many lines to supply sockeye salmon for a growing fresh fish market. The high quality of sockeye salmon annually attracts foreign fleets. Japanese float netters and Soviet factory ships intercept many of these fish within the international waters of the North Pacific.

Beachcombers or pierside naturalists never see the sleek and silvery sockeye salmon alive, and must settle for a glimpse of a recently dead one when a sportsfisherman comes alongside a wharf, a commercial vessel heaves to, or in a fish market.

Although the sockeye's bright red flesh is world renowned, it attracts the most attention from seafood lovers in the Pacific Northwest. A little piece of sockeye on a dish is great; a lot is gourmet!

22 Chinook Salmon

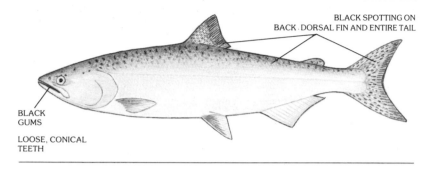

BLACK SPOTTING ON
BACK, DORSAL FIN AND ENTIRE TAIL

BLACK
GUMS

LOOSE, CONICAL
TEETH

SPECIES: ***Oncorhynchus tshawytscha***—from the Greek *onkos* and *rhynchos*, meaning "hook" and "nose;" and *tshawytscha*, the Russian name for this fish in Kamchatka, USSR.

ALTERNATE NAMES: spring salmon, king salmon, blackmouth salmon, quinnat salmon, chub salmon, Sacramento River salmon, Columbia River salmon, chinook, spring, king, quinnat, hookbill, tyee for specimens of 14 kg (30 pounds) and over.

MAXIMUM RECORDED SIZE: 147 cm (58 inches) and 57.3 kg (126 pounds).

DISTRIBUTION: San Diego, southern California, to the Bering Sea, and Coppermine in the Canadian Arctic; along the Asian Pacific coast from the Anadyr River, USSR to Hokkaido northern Japan. The many worldwide transplants attempted have as yet only established viable, reproducing populations in the Great Lakes, New Zealand, and Chile.

Trollers and moochers eagerly seek the mighty chinook salmon, the acknowledged "heavyweight champ" of the Pacific Northwest, and using herring either whole or as "strip," or artificial lures during dawn or dusk hours, may catch trophy-sized specimens. Small, precocious male chinook salmon, called "jacks" or "jack springs," as well as moderate-sized specimens bite readily throughout daylight hours.

Occasionally divers may sight a chinook salmon cruising above them in the clear green water.

Using numerous hook and line rigs, trollers harvest many chinook salmon for the fresh fish trade, while gillnetters and seiners supply their catch for canning or freezing. An ever increasing number of major hatcheries propagate this valuable species, both in Canada and the USA.

Rarely do shorebound naturalists see the fast-swimming chinook salmon as an adult, and they can see it only slightly more often as a juvenile.

Whether "red" or "white," chinook salmon flesh is excellent eating. Try stuffing a whole fish with your favourite dressing and then barbecuing or baking it for a real taste treat. "White spring" reportedly are best for smoking.

23 Coho Salmon

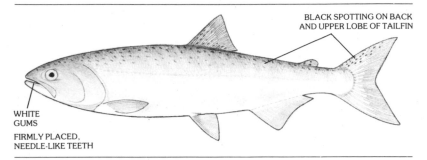

BLACK SPOTTING ON BACK
AND UPPER LOBE OF TAILFIN

WHITE
GUMS

FIRMLY PLACED,
NEEDLE-LIKE TEETH

SPECIES: *Oncorhynchus kisutch* —from the Greek *onkos* and *rhynchos*, meaning "hook" and "nose;" and *kisutch* a Russian name for the fish in Kamchatka, USSR.

ALTERNATE NAMES: silver salmon, coho, silver, hooknose, blueback (young only), silversides, dog salmon*, jack salmon*, sea trout*.

MAXIMUM RECORDED SIZE: 98 cm (38.5 inches) and 14 kg (31 pounds).

DISTRIBUTION: Punta Chanalu Bay, northern Baja California, Mexico to the Bering Sea and the Aleutian Islands; along the Pacific Asian coast from the Anadyr River, USSR. to northern Korea and northern Honshu Island, Japan. Transplants to Argentina, New Hampshire, Maine, Maryland, Chile, Alberta, Lake Michigan and Lake Superior have as yet only established viable populations in the last four regions.

Aerial acrobatics performed by the battling coho salmon give it top billing as a Pacific Northwest gamefish and any of a wide variety of tackle and techniques, such as flies, jigs, plugs, spoons, jack herring or herring strip when trolling near the surface, live herring while mooching, or herring strip for strip casting, may be used to catch this aggressive fish. Early in coho season, April to June, "bluebacks"—young specimens of only a few kilograms—bite readily, but these same fish, if released, will grow up to 10 kg by August or September the same year.

During the summer, divers who look up into the green water above may see the silhouettes of adult coho salmon cruising by, particularly if herring are schooling in the area.

Towing many lines, each with numerous hooks, trollers take coho salmon and sell them fresh while gillnetters and purse seiners trap large numbers for canning, smoking or freezing.

In spring and summer, young orange-tinged, parr-marked coho salmon dart quickly through shallow water near shore, or around jetties and piers.

Bright red flesh with a rich flavour make the coho salmon a seafood lover's dream. Coat a steak in egg and breadcrumbs with a touch of tarragon and fry in butter.

*—incorrect

36

24 Steelhead Trout

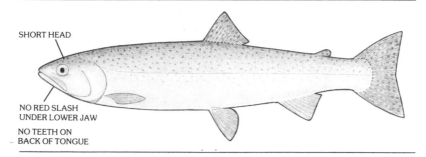

SHORT HEAD

NO RED SLASH
UNDER LOWER JAW

NO TEETH ON
BACK OF TONGUE

SPECIES: *Salmo gairdneri*—from the Latin *salmo*, meaning "salmon;" and honouring Dr. Meredith Gairdner, an early naturalist.

ALTERNATE NAMES: coastal rainbow trout, rainbow trout, steelhead, hardhead, metalhead, ironhead, halfpounder, Gairdner's salmon, salmon trout.

MAXIMUM RECORDED SIZE: 114 cm (45 inches) and 19.5 kg (43 pounds).

DISTRIBUTION: northern Baja California, Mexico, to Bristol Bay on the Bering Sea coast of Alaska and the Aleutian chain. The rainbow trout, of which the steelhead is a sea-run race, has been successfully transplanted throughout the world.

While anglers fishing in salt water rarely expect to catch a steelhead trout, they do take the occasional specimen with "bobbers" or salmon spoons; there is an immense volume of literature and folklore, however, on the joys of "steelheading" in rivers and streams, where legions of anglers avidly pursue this legendary fighter. To enhance this world-famous fish, governments annually spend vast sums on hatcheries and habitat protection.

Only divers frequenting clear rivers and estuaries might perhaps see the sleek, fast steelhead trout; usually, though, turbid water drastically decreases sighting possibilities.

An ever-increasing rift widens between sport anglers and commercial fishermen over the valuable steelhead trout. While gillnetting, purse seining, or trolling for the various Pacific salmon, professional harvesters accidentally take varying quantities of this trout while the covetous angler heartily condemns any such exploitation. This disputed catch is canned for market.

The swift open-ocean steelhead trout is only accessible to shorebound naturalists when it moves into spawning streams where, unfortunately, turbid water often obscures it.

Seafood lovers extol the delights of feasting on steelhead trout; its rich flesh is particularly tempting when baked or barbecued. Try canning some too!

25 Coastal Cutthroat Trout

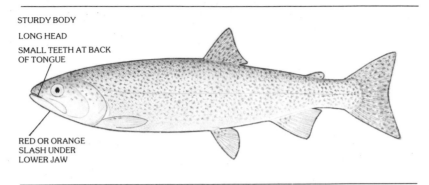

STURDY BODY

LONG HEAD

SMALL TEETH AT BACK
OF TONGUE

RED OR ORANGE
SLASH UNDER
LOWER JAW

SPECIES: *Salmo clarki clarki*—from the Latin *salmo*, meaning "salmon;" and honouring William Clark of the famous Lewis and Clark Expedition.

ALTERNATE NAMES: coastal cut-throat trout, coast cutthroat trout, red-throated trout, Clark's trout, sea trout, coastal cutthroat.

MAXIMUM RECORDED SIZE: 76 cm (30 inches) and 7.8 kg (17 pounds).

DISTRIBUTION: Eel River, northern California, to Prince William Sound, Gulf of Alaska. (Another race of this fish, *Salmo clarki lewisi*, inhabits adjacent interior fresh water haunts not connected with the sea.)

Lurking along shallow, gravelly beaches, the coastal cutthroat trout readily strikes at artificial lures cast from shore or towed from a boat; the proximity of a small stream often further enhances angling chances. Fly fishing enthusiasts who cast in early, mist-shrouded mornings may actually stalk this quarry, frequently seeing it swirl the calm surface water. The best flies simulate the coastal cutthroat trout's primary spring prey—migrating chum salmon fry, pink salmon fry, or threespine stickleback. This active, battling lightweight gamefish ascends small streams in late winter or early spring to spawn.

The shallow beach or estuarine habitat of the coastal cutthroat trout is often too murky for a diver to view this swift and wary fish.

No commercial fishery exists for the small coastal cutthroat trout and few professional fishermen venture near enough to shore to take this fish.

During low-water summer months look closely around sunken logs or along shaded estuarine banks for the lurking coastal cutthroat trout. Keen-eyed beachcombers may occasionally notice a feeding specimen ripple calm, shallow water.

The pinkish flesh of the coastal cutthroat is particularly moist and tender—an excellent breakfast in the outdoors. Indoors too!

26 Dolly Varden

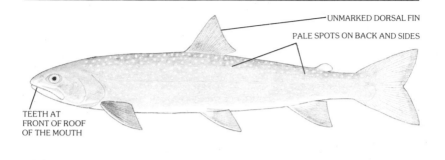

UNMARKED DORSAL FIN

PALE SPOTS ON BACK AND SIDES

TEETH AT
FRONT OF ROOF
OF THE MOUTH

SPECIES: *Salvelinus malma* —from the old European term *salvelinus* for "char;" and *malma*, a Russian name for the fish in Kamchatka, USSR.

ALTERNATE NAMES: dolly varden char, dolly varden charr, red spotted char, Pacific brook char, western brook char, sea char, dolly, bull trout*, red spotted trout*, dolly varden trout*, Rocky mountain red spotted trout*, sea trout*, salmon trout*, brook trout*.

MAXIMUM RECORDED SIZE: 127 cm (50 inches) and 18.3 kg (40 pounds).

DISTRIBUTION: McCloud River, northern California, to the Seward Peninsula, Bering Sea coast of Alaska and throughout the Aleutian Islands; along the Asian Pacific coast from tne Anadyr River, USSR, to Hokkaido, northern Japan and the Yalu River, Korea.

Historically scorned as a scourge upon trout and salmon, the dolly varden—honouring a Charles Dickens character who wore colourful clothes—once was the subject of bounties. While this underrated gamefish does feed upon salmon and trout and their eggs, it also preys upon capelin, smelts, sticklebacks, sculpins, shrimps, insects, clams, and worms. In addition to those natural baits, the estuarine dolly varden takes brightly coloured spoons, spinners and flies, if cast and then retrieved at moderate speeds. Although not renowned as a mighty scrapper, this char will sometimes surprise with its tenacity.

Only the diver who swims in clear estuaries is likely to notice the dolly varden.

No commercial dolly varden fishery exists and incidental catches seldom occur.

If the water is clear enough, the shore-strolling naturalist might see the sluggish dolly varden in gravelly estuaries, particularly in late summer when the various salmon are migrating to spawn.

Similar to trout, the dolly varden is delicious, particularly when pan-fried and fresh. Poach in white wine.

*—incorrect

THE CODFISHES
(Family: Gadidae)

Along with the true cods, the tomcods, pollacks, hakes, cusks, ling, rockling, haddocks, whitings and the burbot comprise the family of fishes scientifically termed Gadidae. Of the nearly sixty recognized living species, all but one, the lake- or river-dwelling burbot, live in salt water and only a few others may wander into brackish environments from their shallow coastal or deeper continental shelf habitats. Gregarious by nature, most gadids will form schools often of considerable size, when cruising near the surface, along the bottom, or through all various midwater depths between. The temperate and boreal waters throughout all seas of the northern hemisphere supply living space for most codfishes including the European ling, the world's largest at 2 m (79 inches); very few varieties inhabit similar southern hemisphere haunts.

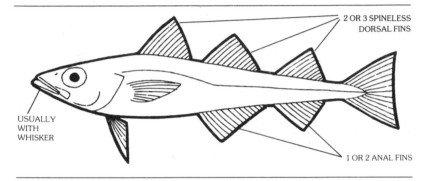

2 OR 3 SPINELESS DORSAL FINS

USUALLY WITH WHISKER

1 OR 2 ANAL FINS

Prehistoric relatives of present-day gadids apparently thrived in the marine environment also, as 30-million-year-old mid-Oligocene fossils indicate. Three Pacific Northwest fish families ally closest to the codfishes: the elongate eelpouts, the secretive brotulas, and the abyssal and seldom-seen rattails or grenadiers with their long whip-like tails. None have body spines or heavy spine-supported fins, but all have pelvic fins located far forward on the undersides of their bodies, often just below their heads.

The renowned codfishes, most species of which may grow to excellent marketable sizes, rank economically among the most important fishes in the world. In its constant quest for new and virgin jigging grounds, the European cod fishery significantly spurred historic western exploration. Now all manner of fishing techniques—from traditional hand-lining and long-lining aboard open dories to the modern seining and trawling techniques utilized by large, fast ships—annually take vast quantities of gadids for processing into numerous products such as fresh or frozen fillets, the ever-popular salt cod, and finnan haddie.

27 Pacific Cod

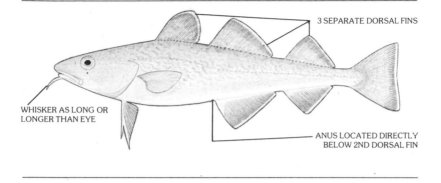

3 SEPARATE DORSAL FINS

WHISKER AS LONG OR
LONGER THAN EYE

ANUS LOCATED DIRECTLY
BELOW 2ND DORSAL FIN

SPECIES: *Gadus macrocephalus* —from the Latin *gadus*, meaning "codfish"; and the Greek *macro*, and *cephalus*, "big," and "head."

ALTERNATE NAMES: grey cod, gray cod, true cod, Alaska cod, whisker cod, greyfish.

MAXIMUM RECORDED SIZE: 117 cm (46 inches) and 22.7 kg (50 pounds).

DISTRIBUTION: Santa Monica, southern California, to St. Lawrence Island at the Bering Straits, throughout the Bering Sea, and south to Darien, Korea, as well as Tottori and Oshu, Japan.

Fish for Pacific cod from wharves, jetties, or boats over sandy and muddy bottoms. Although baits such as herring, marine worms, clams, or shrimps most often tempt this species, lures and jigs may also prove effective. Particularly when hooked on light tackle and in shallow water, a large Pacific cod often fights tenaciously for its freedom.

While swimming over soft bottoms, divers sometimes encounter a Pacific cod cruising alone or as part of a school. This gray, whiskered fish usually swims along just off the bottom in search of prey such as young walleye pollock, pricklebacks, eelpouts, octopus, brittle stars, snails, and crabs. During the winter spawning season a large female Pacific cod may swell with as many as 6,400,000 tiny eggs which she will release for fertilization before she abandons them and lets them float on ocean currents.

Though North Americans harvest the Pacific cod in significant quantities, Asians prize it even more and their fleets outharvest ours annually. Historically, hook and lining dominated the North American fishery, but later the modern net-hauling vessels arrived and now processing plants aboard factory ships fillet the catch or render it into fish sticks.

Sometimes dockside observers see Pacific cod swimming in clear water around jetties, pilings, or particularly near bait-herring ponds.

When gutted, then chilled quickly, the Pacific cod is one of the finest of eating fish. Try it deep-fried for fish and chips. Chowder fans save the head!

28 Pacific Tomcod

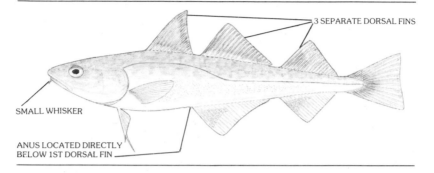

3 SEPARATE DORSAL FINS

SMALL WHISKER

ANUS LOCATED DIRECTLY
BELOW 1ST DORSAL FIN

SPECIES: *Microgadus proximus*—from the Greek *micros* and *gadus*, meaning "small" and "codfish;" and the Latin *proximus*, "next."

ALTERNATE NAMES: tomcod, wachna.

MAXIMUM RECORDED SIZE: 30.5 cm (12 inches).

DISTRIBUTION: Point Sal, central California, to Unalaska Island and perhaps the Bering Sea.

When using shrimps, small crabs, marine worms, clams, or pieces of fish on smallish hooks, anglers often catch the Pacific tomcod—itself an excellent bait for large quarry. Fishing for this small codfish is most rewarding when done from piers, wharves or boats located over sandy or muddy bottoms.

Easily approached during shallow-water night dives, schooling juvenile Pacific tomcod are particularly abundant during summer and autumn evenings. These young specimens frequently swarm with juveniles of the larger-eyed walleye pollock, or even occasionally with the longer-whiskered Pacific cod, to feed upon shrimp-like planktonic creatures. Much less frequently, though, does the aquanaut find the deeper-dwelling adult Pacific tomcod.

Of marginal economic value because of its small size, the Pacific tomcod nevertheless commonly shows up in fine-meshed groundfish or shrimp trawls and knowledgeable commercial fishermen may actually select this fish from their catches for personal consumption. The dusky-coloured Pacific tomcod spawns in late winter or spring and by June the newly hatched, transparent larvae begin a search for tiny planktonic prey at or near the surface.

While juvenile Pacific tomcod are schooling in shallow bays or inlets during summer and autumn, dockside naturalists have the best chance to see them with the aid of lights after dark. Distinguishing them from the young walleye pollock, though, proves most difficult from an above-water vantage point.

The Pacific tomcod's sweet and delicate flavour and fine texture make it excellent eating. No need to fillet this small fish—cook it whole because the flesh will fall easily from the backbone.

29 Walleye Pollock

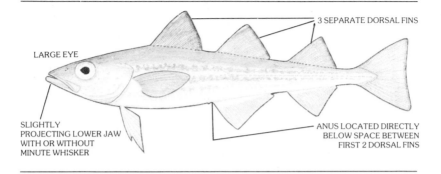

3 SEPARATE DORSAL FINS

LARGE EYE

SLIGHTLY
PROJECTING LOWER JAW
WITH OR WITHOUT
MINUTE WHISKER

ANUS LOCATED DIRECTLY
BELOW SPACE BETWEEN
FIRST 2 DORSAL FINS

SPECIES: *Theragra chalcogramma*—from the Greek *ther*, *agra*, *chalcos*, and *gramma*, meaning "beast," "food for fur seals," "brass," and "mark."

ALTERNATE NAMES: wall-eye pollock, Pacific pollock, Alaska pollack, Alaska pollock, bigeye, whiting, bugeye, walleye, scrapcod.

MAXIMUM RECORDED SIZE: 91 cm (36 inches).

DISTRIBUTION: Carmel, central California, to St. Lawrence Island on the Bering Straits, throughout the Bering Sea and south to Sagami Bay, southern Japan, and Korea.

While the schooling walleye pollock often pursues lures and jigs, it most readily bites at baits such as marine worms, shrimps or herring. Living in shallow to moderate depths and generally not too far from a sandy or muddy bottom, it is frequently caught from boats and jetties. Do not expect great sport from this slightly-built fish.

Although divers seldom see the adult walleye pollock, they frequently encounter the active juvenile because it often forms mixed schools with the smaller-eyed Pacific tomcod just off sandy or muddy bottoms. For best viewing of the walleye pollock, dive on a summer or autumn night. A specimen can be easily transfixed for closeup study with the beam of your underwater flashlight. Foraging on such prey as shrimps, sand lance, and herring, the walleye pollock itself falls prey to sea birds, marine mammals and other large fish.

Increasingly important North American trawl fisheries for walleye pollock still lag far behind those in Korea, in the USSR, and in Japan, where even the roe, *tarako*, is highly prized. While historically underutilized—mostly for mink feed—the walleye pollock is ever increasing in importance as food for people in the Pacific Northwest.

On a calm summer or autumn evening, the pierside naturalist may often see young walleye pollock darting amid beams of light shining into the water.

To avoid the numerous bones, fillet fresh walleye pollock before chilling the firm sweet flesh. Braise fillets in white wine.

30 Pacific Hake

LARGE SCALES ON BODY

2 DORSAL FINS WITH THE 2ND LONG AND DEEPLY NOTCHED

PROTRUDING LOWER JAW WITH NO WHISKER

LONG ANAL FIN, DEEPLY NOTCHED

SPECIES: *Merluccius productus*—from the word *merluccius*, meaning "sea pike;" and the Latin *productus*, "drawn out."

ALTERNATE NAMES: Pacific whiting, California hake, hake.

MAXIMUM RECORDED SIZE: 91 cm (36 inches).

DISTRIBUTION: an isolated population in the Gulf of California; and from Bahia Magdalena, southern Baja California, Mexico, to the Gulf of Alaska and the Asian coast.

Stillfishing salmon anglers, or "moochers" who use herring as bait periodically catch the schooling Pacific hake as an unwanted incidental, particularly in summer when darkness falls and this sharp-toothed fish rises to feed near the surface. Although a few wharf-bound enthusiasts avidly fish for it, most anglers find little excitement in hooking the sluggish Pacific hake.

Although the elongate Pacific hake swims through deep bottom haunts by day, it usually rises to invade diveable shallow or surface waters after dusk. On summer or autumn nights particularly, look for this slender silvery form.

Formed by day in long narrow schools at depths as great as 980 m (3,280 feet), Pacific hake aggregations rise to the surface, then disperse to feed independently upon tiny shrimp-like plankton after dark. By daybreak, though, these mobile fishes have again congregated in deep dark waters. Historically viewed as an expensive nuisance that clogged salmon gillnets and groundfish trawls, the Pacific hake is slowly, though fitfully, being established as a valuable commercial species. A growing foreign market, together with concerted research into the refrigeration of its easily spoiled flesh, promises a great future for this new, largely untapped fishery.

Where good artificial lighting exists, a wharf-bound naturalist may notice the nocturnally foraging Pacific hake because it occasionally swims into shallow inlets or bays.

Promptly filleted and chilled, Pacific hake is fine eating *au gratin*; do not be discouraged by some brownish colour on the long, slender fillets.

44

THE BROTULAS
(Family: Ophidiidae)

Over two hundred known living species of brotulas and cusk-eels form a very large family scientifically called the Ophidiidae, Greek for "snake-like family." Most species colonize tropical marine haunts where they hide within deep caves and crevices among the shallow coral reefs, particularly in the vast Indo-Pacific. Unusual and diverse habitats, such as caverns very deep on the ocean floor or on lake bottoms, harbour a few specialized ophidiids. For example, the Sunda trench in the south Pacific, 6.6 km (4 miles) down from the surface, is home for the world's deepest dwelling fish species, a brotula. Relatively few ophidiids invade cooler temperate seas, and apparently only one, the red brotula, lives along Pacific Northwest shores.

The total darkness of deep marine or fresh water habitats apparently stimulated evolution of certain features in abyssal brotulas. Many of these secretive ophidiids now no longer possess functional eyes, while others may have only very tiny rudimentary ones. Compensating for this partial or total loss of sight, these fish have special sensors, similar to taste buds, located on their long, tapering and paired pelvic fins, and those sensors allow these slow swimmers to grope for their food before actually devouring it.

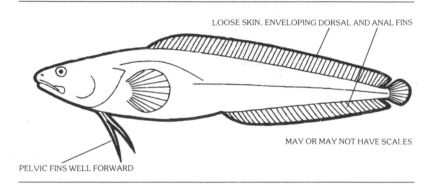

LOOSE SKIN, ENVELOPING DORSAL AND ANAL FINS

MAY OR MAY NOT HAVE SCALES

PELVIC FINS WELL FORWARD

As far back as 63 million years, during the Paleocene epoch, ancestral brotulas left impressions or remains behind. Although modern ophidiids primarily huddle in tropical domains, two families most closely related to them, the codfishes and eelpouts—members of all three groups lack body spines and fins supported by spiny rays—are primarily cool water creatures. Within subsistence-level native cultures of many tropical islands, the crevice-loving brotulas, some of which attain 1.2 m (4 feet), provide an excellent source of food. However, fast, modern fishing interests, with their emphasis on the quality, quantity, and accessibility of exploitable species, find ophidiids not worth harvesting. A second Pacific Northwest species, the giant cusk-eel, dwells at abyssal depths.

31 Red Brotula

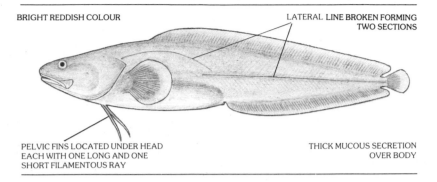

BRIGHT REDDISH COLOUR

LATERAL LINE BROKEN FORMING TWO SECTIONS

PELVIC FINS LOCATED UNDER HEAD EACH WITH ONE LONG AND ONE SHORT FILAMENTOUS RAY

THICK MUCOUS SECRETION OVER BODY

SPECIES: *Brosmophycis marginata*—from *brosome*, an established name for a similar fish, the cusk; with the Greek *phycis* meaning "like'; and the Latin *marginata*, "edge."

ALTERNATE NAMES: red brotulid.

MAXIMUM RECORDED SIZE: 46 cm (18 inches).

DISTRIBUTION: Ensenada, northern Baja California, Mexico, to Petersburg, southeastern Alaska.

Chances of catching a red brotula are slim because of its secretive, cave-dwelling nature. In fact, to catch one, an angler would have to drop a hooked offering of shrimp or small baitfish practically into the red brotula's deep water cave.

Diligent searching along rocky reefs and cliff faces at depths below 20 m (66 feet) occasionally reveals a red brotula looking out from a deep and rocky cavern. Sometimes, though, a surprised aquanaut may even detect a mated pair hovering at the entrance to a cave, or backing into it when confronted. A colourful find, particularly appealing to the underwater photographer.

The elusive red brotula is of no direct economic value because fishermen seldom catch it. In certain localized areas an occasional specimen wanders into a prawn trap to dine upon the captives, and even less often a shrimp dragger incidentally nets a red brotula temporarily out of its crevice or cavern at depths to 256 m (840 feet).

Only through extremely unusual circumstances would a beachcomber or wharf-bound naturalist encounter the deepwater red brotula.

A sweet, fine-textured flesh—a gourmet's delight. Unfortunately, neither a gourmet nor anyone else has much opportunity to eat it. If lucky enough to obtain a red brotula, wash its scaleless body thoroughly to remove the thick red slime.

THE EELPOUTS
(Family: Zoarcidae)

The Zoarcidae, translating literally from the Greek to mean "maintaining life family," is a moderately large group containing nearly sixty-five recognized living species of eelpouts, soft pouts, slipskins, and ocean pouts. Without exception all varieties live in salt water, from shallow inshore habitats to the abyssal regions of oceanic trenches and at various depths between. Although a few slim zoarcids hover in mid-water well off the bottom, most settle right on soft, gently sloping substrates of the sea floor. An intriguing group geographically, the elongate eelpouts thrive primarily in boreal and temperate latitudes of both hemispheres; the tropics have almost none. Those few tropical forms, like their northern counterparts, live in cool or cold habitats—the deep, dark haunts of equatorial regions.

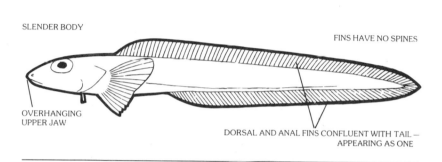

SLENDER BODY

FINS HAVE NO SPINES

OVERHANGING
UPPER JAW

DORSAL AND ANAL FINS CONFLUENT WITH TAIL —
APPEARING AS ONE

Although only a few fossilized eelpout otoliths, or inner ear bones, have as yet been unearthed, this scant prehistoric record indicates that zoarcids lived as long ago as one million years, during the Pliocene epoch. After comparing eelpouts, brotulas, and codfishes, ichthyologists believe them to be closely related; all member species lack body or fin spines and have pelvic fins located far forward on their undersides. Aside from the large ocean pout, which attains a length of 107 cm (42 inches), few eelpouts grow large enough for commercial exploitation as human food; their immense, indirect worth as food for economically significant bottomfishes, however, defies calculation.

A large array of Pacific Northwest zoarcids inhabit regions that are very deep and inaccessible to most people: the bigfin eelpout, the twoline eelpout, the soft eelpout, the Alaska eelpout, the longsnout eelpout, the snakehead eelpout, the shortjaw eelpout, the deepwater slipskin, the blackmouth slipskin, the pallid slipskin, the stout slipskin, the stubraker slipskin, and the Pacific softpout. One other species with a primarily northern distribution is the wattled eelpout, possessing a prominent flap beneath its lower jaw.

32 Blackbelly Eelpout

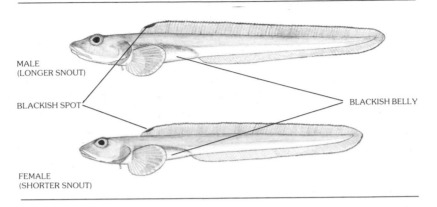

MALE
(LONGER SNOUT)

BLACKISH SPOT

BLACKISH BELLY

FEMALE
(SHORTER SNOUT)

SPECIES: *Lycodopsis pacifica*—from the Greek *lycodes* and *opsis*, meaning "wolffish" and "like;" and emphasizing the Pacific Ocean.

ALTERNATE NAMES: black-bellied eelpout.

MAXIMUM RECORDED SIZE: 46 cm (18 inches).

DISTRIBUTION: Ensenada, northern Baja California, Mexico, to Afognak and perhaps Unalaska Island, in the Gulf of Alaska.

A small fish lurking upon muddy or sandy substrates, the small-mouthed blackbelly eelpout rarely sees baited hooks, let alone bites at them. Particularly if used live as bait, this fish attracts large rockfishes, greenlings or codfishes.

At night, the very common blackbelly eelpout moves into shallow, diveable depths to feed on the small clams, worms, shrimps, and brittle stars which share its muddy- or sandy-bottomed habitat. Search at depths below 15 m (50 feet) to find this slender creature, sometimes coiled in a snake-like fashion. A mature female, recognized by her shorter and smaller head, may lay as many as 52 eggs during the August to January breeding season. During some future underwater exploration, try to confirm whether blackbelly eelpout parents guard their eggs or not.

Often entwining themselves in the webbing, vast quantities of these economically insignificant creatures fill fine-meshed shrimp trawls dragged over soft, level bottoms at depths to 400 m (1,320 feet). Barring death in such nets or in other ways, the males—which usually outlive the females—may attain at least five years of age.

The pale blackbelly eelpout rests upon silty or sandy bottoms at depths too great for dockside or shoreline naturalists to see.

The flesh tastes good but the thin and elongate bodies of the blackbelly eelpout stir up little enthusiasm among seafood fanciers. Possibly try it as a bouillabaisse ingredient.

48

33 Black Eelpout

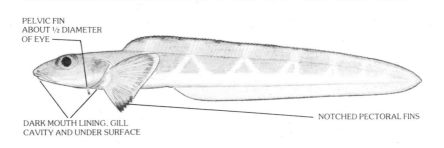

PELVIC FIN
ABOUT ½ DIAMETER
OF EYE

DARK MOUTH LINING, GILL
CAVITY AND UNDER SURFACE

NOTCHED PECTORAL FINS

SPECIES: *Lycodes diapterus* — from the Greek *lycodes*, *di*, and *apteros*, meaning "wolffish," "two," and "fin."

ALTERNATE NAMES: blackfin eelpout, black-finned eelpout.

MAXIMUM RECORDED SIZE: 33 cm (13 inches).

DISTRIBUTION: San Diego, southern California, to the southeastern coast of the Bering Sea, Alaska and through the Aleutian chain, to the Bering Sea coast of the USSR and northern Japan.

 Very few anglers lower tiny baited hooks to depths below 50 m (165 feet) — consequently the black eelpout, a denizen of sandy or muddy bottoms, is probably never caught. It could be a very useful bait, however, if an enterprising angler could collect enough specimens.

 Because 50 m stands as the minimum depth record for the black eelpout, on your next dive do not expect to come at it facemask to snout. Some member of the underwater fraternity, however, may yet discover a shallow water black eelpout population. Such novel sightings certainly add variety to diving experiences and provide valuable contributions to our knowledge of natural history.

 Although no commercial black eelpout fishery exists, shrimp trawlers dragging their nets over soft, level bottoms at depths between 50 and 2,260 m (165 and 7,460 feet), frequently take this dark-coloured, often striped species. After removing entwined specimens from a trawl net, notice the bright blue often present on the pectoral fins, a coloration common to several other deep-dwelling bottomfishes of the Pacific Northwest, the blackfin poacher for example.

 Beachcombers and dockside strollers are effectively isolated from the deepwater haunts of the black eelpout.

Although firm and tasty of flesh, the elongate but small black eelpout inspires little culinary interest — unless, of course, one is very, very hungry.

34 Shortfin Eelpout

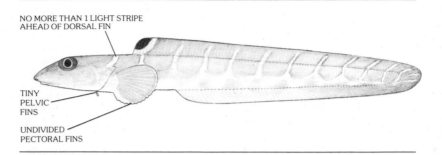

NO MORE THAN 1 LIGHT STRIPE
AHEAD OF DORSAL FIN

TINY
PELVIC
FINS

UNDIVIDED
PECTORAL FINS

SPECIES: *Lycodes brevipes* —from the Greek *lycodes*, meaning "wolffish;" as well as the Latin *brevis* and *pes*, "short" and "foot."

ALTERNATE NAMES: none.

MAXIMUM RECORDED SIZE: 30 cm (11.8 inches).

DISTRIBUTION: Oregon to the Bering Sea coast of Alaska, through the Aleutian chain to the Okhotsk Sea, USSR.

 That a bottom-fishing angler, even one using tiny baited hooks over a muddy or sandy bottom, would take the small-mouthed shortfin eelpout is extremely unlikely. After such a rare event, though, the lucky angler should retain the specimen for bait because one of its known predators is the popular lingcod.

 One specimen of the shortfin eelpout was taken in a trawl net from English Bay at a depth of less than 15 m (50 feet) and photographed—the depth being a record indicating that SCUBA enthusiasts might perhaps sight this species as it rests upon a sandy or muddy substrate. Because this pale grayish fish often huddles near shelter, it may be difficult to spot.

 While harvesting valuable shrimp species, trawlers dragging fine meshed nets over smooth, soft bottoms at depths down to 642 m (2,219 feet) sometimes capture the slender shortfin eelpout. This worthless catch, representing only time wasted in sorting the take, is disdainfully thrown overboard along with other creatures. Only indirectly, as food for economically valuable groundfishes, is the shortfin eelpout of any value to the commercial fisherman.

 Considering the minimum depth record for the seldom-seen shortfin eelpout, the surface-bound observer is most unlikely ever to see it.

With any number of tasty larger marine fishes readily available to the Pacific Northwest gourmet, the small and seldom seen shortfin eelpout holds little culinary appeal.

THE STICKLEBACKS
(Family: Gasterosteidae)

The sticklebacks, or Gasterosteidae, are a small family of fishes containing approximately a dozen living recognizable species, only one of which lives along Pacific Northwest shores. All member species live only in the temperate and boreal waters of the northern hemisphere, spread fairly evenly around the globe. Some of the world's most adaptable fishes are sticklebacks: for example, the familiar threespine stickleback tolerates—and actually flourishes in—fresh water or salt water and all concentrations of brackish in between. Active swimmers that often congregate in large and loosely organized groups, these bony-plated gasterosteid fishes usually frequent very shallow habitats where they can remain at the surface but not far from the bottom.

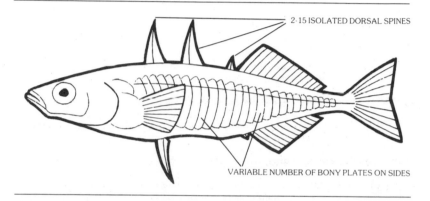

2-15 ISOLATED DORSAL SPINES

VARIABLE NUMBER OF BONY PLATES ON SIDES

Extensive collections by many diligent paleontologists contain vast stores of gasterosteid fossils, and from these ancient, hardened remains it is known that sticklebacks lived as far back as the Miocene epoch, 12 million years ago. Today, the tubeshouts and the pipefishes, two families with Pacific Northwest species, remain as the sticklebacks' closest relatives; members of all three have tube-like snouts with tiny mouths and nearly always have modified scales forming armour-like body plates.

The hardy gasterosteids, few of which grow larger than 20 cm (8 inches), are very popular experimental animals because they are easily and conveniently captured, then easily maintained in aquaria. Innumerable laboratory studies detail the very complex behaviour patterns of these spiny-backed creatures during their courting, mating and territorial interactions. Scientists attach particular relevance to this laboratory work, for it compares most favourably with ecology field studies concerning these animals and their environment. All of these factors combine to make gasterosteid fishes important organisms for the study of pollution, one of the twentieth century's most ominous threats.

35 Threespine Stickleback

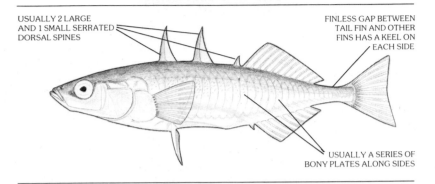

USUALLY 2 LARGE AND 1 SMALL SERRATED DORSAL SPINES

FINLESS GAP BETWEEN TAIL FIN AND OTHER FINS HAS A KEEL ON EACH SIDE

USUALLY A SERIES OF BONY PLATES ALONG SIDES

SPECIES: *Gasterosteus aculeatus*—from the Greek *gastero* and *steus*, meaning "belly" and "bone;" and the Latin *aculeatus*, "spines."

ALTERNATE NAMES: three-spine stickleback, three-spined stickleback, two-spine stickleback, saw-finned stickleback, common stickleback, eastern stickleback, New York stickleback, European stickleback, banstickle, spantickle, tiddler, pinfish[*].

MAXIMUM RECORDED SIZE: 10 cm (4 inches).

DISTRIBUTION: Throughout the fresh, brackish and salt waters of the northern hemishpere; in the marine environment of the Pacific from Rio Rossario, northern Baja California, Mexico, to the Bering Sea coasts; along the North American Arctic coast to Simpson Lagoon and along the Asian shore to southern Japan and Korea.

Too small for capture by conventional angling gear, the abundant and readily available threespine stickleback offers some potential as a baitfish.

While snorkeling to or from a dive site, or while around pilings and piers, look for congregations of silvery threespine stickleback at the surface; rarely will it be sighted, though, below depths of a few metres.

The tiny threespine stickleback effectively evades all commercially-set gear and is of no direct economic value. Indirectly, though, in lakes and streams, this species competes with young salmon for food and itself is eaten by larger marine fishes, sea birds, seals, and even garter snakes.

Both salt and brackish water tidepools harbour the threespine stickleback, particularly while it stakes out territory and breeds. Large numbers of these fishes gather around jetties and wharves; watch a single specimen as it swims jerkily by its pectoral fin strokes, in a manner very distinct from that of other small, silvery fishes. Boat-bound naturalists, far from shore, may even find the surface-inhabiting threespine stickleback hovering beneath bits of flotsam.

Northern natives sometimes freeze threespine stickleback for dog food, but nobody rates it highly as human fare.

[*]—incorrect

THE TUBESNOUTS
(Family: Aulorhynchidae)

Containing only two known living species, the tubesnouts, scientifically called the Aulorhynchidae, are obviously a minute family. Its two equal-sized members are strictly marine and reside within the North Pacific region, one in the shallow coastal waters of Korea and Japan and the other in similar Pacific Northwest habitats.

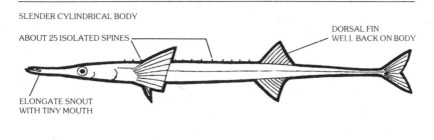

Only a few fossil tubesnouts rest in paleontological collections but dating this sparse aulorhynchid selection shows that ancestors of the two present-day species lived as far back as 35 million years ago, in Eocene seas. Such a very close relationship exists between present-day aulorhynchids and another northern family, the sticklebacks, that, anatomically at least, a tubesnout is essentially a stretched-out, armourless stickleback.

36 Tube-snout

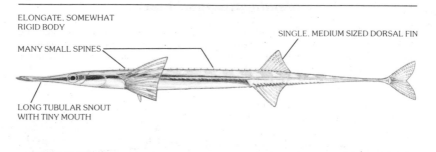

ELONGATE, SOMEWHAT
RIGID BODY

MANY SMALL SPINES

SINGLE, MEDIUM SIZED DORSAL FIN

LONG TUBULAR SNOUT
WITH TINY MOUTH

SPECIES: *Aulorhynchus flavidus*—from the Greek *aulos* and *rhynchos*, meaning "tube" and "nose;" and the Latin *flavidus*, "yellow."

ALTERNATE NAMES: tubesnout, tubenose, needlefish*.

MAXIMUM RECORDED SIZE: 17.9 cm. (7 inches).

DISTRIBUTION: Punta Banda, northern Baja California, Mexico, to Sitka, southeastern Alaska.

The bait-fishing angler seeking rockfish, greenling, or flounder should consider dipnetting the abundant, shallow-water tube-snout for bait. Unless snagged on a tiny jig, this small and minute-mouthed fish is never caught with sport tackle.

If diving in shallow, weedy areas or around docks and pilings, look for tube-snout schools hovering above the bottom in less than 10 m (33 feet); they often contain both young and adults. Be still and patient, and perhaps see a specimen lunge at its small, planktonic, shrimp-like prey. A rowing motion from the tube-snout's pectoral fins propel it somewhat jerkily and unevenly.

Easily evading all varieties of commercial fishing gear, the slender, unmarketable tube-snout never appears as an incidental catch.

During the late spring and early summer breeding season, the patient dockside observer may readily see the detailed spawning ritual of the tube-snout. Each darkened, mature male, with his fluorescent snout and red or blue spots, stakes out a territory, often among marine plants hanging from a float or piling. Accompanied by as many as ten mature females, the industrious male binds seaweed together with thread-like strands of his sticky genital fluid, and waits while successive consorts parade into this nest, each to deposit clusters of up to 60 amber-coloured eggs. After he fertilizes each clutch as it is laid in his nest, the amazing male vigourously guards the nest until, within three weeks, tiny, transparent, 1-mm tube-snout larvae hatch and swim off in search of food.

Briefly pan-fry floured, whole tube-snouts in butter for a crispy snack!

*—incorrect

54

THE PIPEFISHES
(Family: Syngnathidae)

The Syngnathidae is a large family of fishes containing approximately one hundred and seventy-five recognizable living species of seahorses and pipefishes. The elongate pipefishes are essentially stretched out versions of the familiar seahorses, but without the flexible prehensile tails. Aside from one unusual deep water species, all syngnathids live in shallow seas throughout temperate or tropical latitudes, and only a very few tolerant varieties venture into nearby brackish or freshwater habitats.

Within the aquatic world of pipefishes and seahorses, parental care of offspring takes on a novel twist. After the intricate courtship of a mature pair and their subsequent mating, which includes copulation and internal fertilization, the female syngnathid transfers the viable eggs to a special brood pouch located along the male's lower underside. From this time on, he cares for the tiny developing young, while the unburdened female renews regular foraging activities.

BODY ENCLOSED IN ARMOUR OF ENCIRCLING RINGS MAKING IT RIGID

SINGLE DORSAL FIN

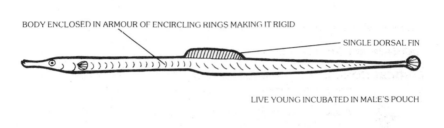

LIVE YOUNG INCUBATED IN MALE'S POUCH

The most ancient recognizable syngnathids apparently lived in early Miocene times, some 50 million years ago. Several present-day fish families, including the sticklebacks, as well as the tropical trumpetfishes and coronetfishes, are the closest surviving relatives of the pipefishes. Tube-like snouts with tiny mouths and swim bladders that do not connect with the gut represent common characteristics of all four groups. The tiny, bony but light-weight carcasses of the various pipefishes, all of which grow no longer than 46 cm (18 inches) in length, stimulate no real fervour among commercial fishermen, but do interest a few groups. For centuries, various oriental cultures have dried these animals for medicinal purposes and an ever-increasing pet industry annually markets countless seahorses and pipefishes to enthusiastic amateur aquarists throughout the world.

37 Bay Pipefish

MALE

TINY MOUTH
NO PELVIC FINS
VERY LONG INFLEXIBLE BODY COVERED WITH ABUTTING BONY PLATES

TINY TAIL

FEMALE

SPECIES: *Syngnathus griseolineatus*—from the Greek *syn* and *gnathos*, meaning "together" and "jaws;" and the Latin *griseo* and *lineatus*, "gray," and "lined."

ALTERNATE NAMES: northern bay pipefish, pipefish, pipe-fish.

MAXIMUM RECORDED SIZE: 33 cm (13 inches).

DISTRIBUTION: Guerrero Negro Lagoon, central Baja California, Mexico, to Sitka, southeastern Alaska.

Anglers never intentionally take the slow-moving bay pipefish, with its tiny tubular snout, but could accidentally snag one; its thin armour-encased body is valueless as bait.

Look very carefully among marine plants, particularly eelgrass, which grow in shallow bays or around pilings to find the slender, well-camouflaged bay pipefish. Watch this creature as it slowly glides in a deliberate search for tiny shrimp-like prey; often carrying its stiff body upright, it moves by vibrating a single transparent dorsal fin. Easily capture one by hand and notice the circular bony rings encasing the creature's body; carefully release it again.

Because nearly all commercial fishing paraphernalia is ineffectual in capturing the unsought, unmarketable bay pipefish, very few are caught, even incidentally.

A shallow water denizen that usually lurks among marine plants, hanging from wharves, or growing along shorelines, or even floating freely in tide lines, the bay pipefish is a common sight for the observant naturalist. In spring, watch for the courting male as he holds his body in an S-shape while nodding or shaking his head at a potential breeding partner. Sometime after her eggs have been internally fertilized, the transient female entwines with her mate and transfers viable ova to his long brood pouch. Weeks later his swollen pouch splits, releasing many tiny, black, thread-like baby pipefish into their permanent environment.

Although some oriental people dry Asian species for medicinal uses, no one actually eats the unappetizing bay pipefish.

THE SANDFISHES
(Family: Trichodontidae)

The Trichodontidae, popularly known as the sandfishes, is a very tiny family of marine fishes containing only two known living species: one lurks along the North Pacific coast of Asia, while the other frequents the shores of western North America, but neither variety enters the freshwater environment. Secretive, sandy bottom dwellers, the trichodontids—no longer than 33 cm (13 inches)—bury themselves in the soft substrate and wait with their upturned, fringed mouths to grasp unsuspecting prey.

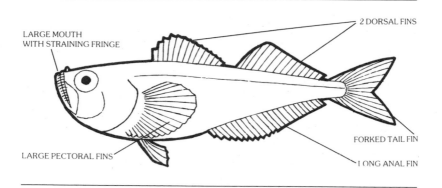

LARGE MOUTH
WITH STRAINING FRINGE

2 DORSAL FINS

LARGE PECTORAL FINS

FORKED TAIL FIN

LONG ANAL FIN

Constant and diligent searching may yet unearth some trichodontid fossils, but at present the group's ancient history remains unknown. Uncertainty also exists regarding the sandfishes' closest present-day relatives although the bottom-dwelling stargazers, with their upturned mouths, may be nearest kin. While no interest shows for the seldom-seen Pacific Northwestern variety, resourceful Asians actively pursue their local species for noteworthy regional fisheries.

38 Pacific Sandfish

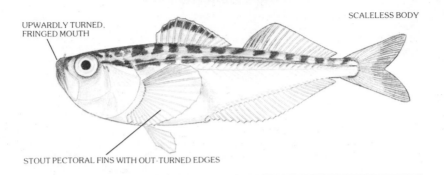

UPWARDLY TURNED, FRINGED MOUTH

SCALELESS BODY

STOUT PECTORAL FINS WITH OUT-TURNED EDGES

SPECIES: *Trichodon trichodon*—from the Greek *trichos* and *odons*; meaning "hair" and "teeth."

ALTERNATE NAMES: sandfish, sand-fish.

MAXIMUM RECORDED SIZE: 30 cm (12 inches).

DISTRIBUTION: San Francisco Bay, central California, to the Bering Sea coast of Alaska, through the Aleutian chain to Kamchatka, USSR.

A surfcaster after bottomfishes along sandy beaches, particularly those beaches directly exposed to the Pacific's waves, could possibly hook a Pacific sandfish. Live bottom-dwelling invertebrate animals would undoubtedly be best as bait for this small fish.

The diver would be unlikely to observe the Pacific sandfish because it lives in a very shallow, turbid environment and spends much of its time completely buried in the sand. Night-time netting studies, though, suggest that this seldom-observed fish prowls actively after dark: perhaps a night dive, in shallow water over a sandy bottom, might produce a sighting.

A ready market does not exist for the Pacific sandfish and no commercial fishermen set nets along its shallow, sandy habitat.

On a visit to a sandy, wave-swept beach, the dauntless beachcomber may literally stumble across the Pacific sandfish, known as *anamlukh* to the Aleut Indians, because the receding tide may have left it buried in the sand. Try dragging your feet as you walk along the edge of the water; perhaps one might wriggle free while you are digging clams. Apparently, the Pacific sandfish spawns in late winter, with mature individuals of at least two years of age participating in the courtship and breeding routine. After hatching, each larva swims and forages at the surface until it attains about 2.5 cm (1 inch) in length, then settles to a sandy bottom.

Copy the Japanese and Korean seafood fanciers and prepare treats from a Pacific sandfish. Sushi anyone?

THE SURFPERCHES
(Family: Embiotocidae)

Twenty-three known living species of surfperches, seaperches and viviparous perches form a small, noteworthy family of fishes, scientifically called the Embiotocidae. With the single exception of the freshwater dwelling California tule perch, embiotocids are all marine creatures and only a few of the others may wander into brackish river estuaries. Shallow, and even intertidal locales along sandy or muddy shores, often adjacent to rocky bottom coasts, provide habitats for most species; few live deeper than 30 m (100 feet). All members, none of which grow longer than 48 cm (19 inches), inhabit the temperate North Pacific: two small species thrive along Korean and Japanese shores; the remainder are North American.

Unlike a majority of fishes, the amazing embiotocids are viviparous because they actually bear large, fully-developed living young. After copulation with, and fertilization by mature breeding males, ripe female surfperches slowly but steadily go through structural as well as functional changes to accommodate as many as forty developing and growing young. For nourishing these babies an elaborate internal system, not entirely unlike that of a female mammal, evolves over the nearly year-long gestation period. When ultimately jettisoned into the sea, the newborn—miniature replicas of their parents—have a special size advantage over other young fish that hatch directly from tiny eggs and swim immediately in the sea.

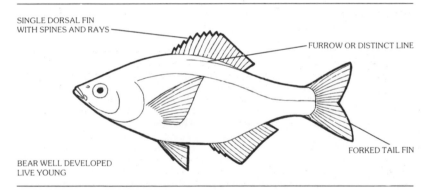

SINGLE DORSAL FIN WITH SPINES AND RAYS

FURROW OR DISTINCT LINE

FORKED TAIL FIN

BEAR WELL DEVELOPED LIVE YOUNG

Numerous paleontological digs in dry Californian locales have unearthed most of the world's known embiotocid fossils. Very similar to present-day varieties, prehistoric surfperches appeared in the Pacific Ocean at least 20 million years ago, during the Miocene epoch. Although they occur in Pacific Northwestern waters, three species are seldom seen: the calico surfperch, with its tall spiny dorsal fin; the walleye surfperch, similar to the fish on page 63, but with black-tipped pelvic fins; and the spotfin surfperch, with its black-splotched, spiny dorsal fin.

39 Striped Seaperch

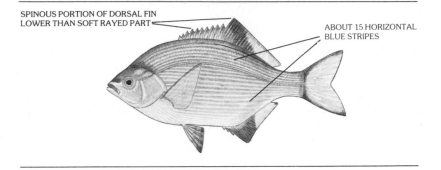

SPINOUS PORTION OF DORSAL FIN
LOWER THAN SOFT RAYED PART

ABOUT 15 HORIZONTAL
BLUE STRIPES

SPECIES: *Embiotoca lateralis*—from the Greek *embios* and *tocos*, meaning "living" and "bring forth;" and the Latin *lateralis* referring to the stripes.

ALTERNATE NAMES: blue seaperch, blue perch.

MAXIMUM RECORDED SIZE: 38 cm (15 inches).

DISTRIBUTION: Point Cabras, northern Baja California, Mexico, to Wrangell, southeastern Alaska.

 Particularly during the summer along shallow, rocky shores or around wharves and pilings, the striped seaperch is good quarry for the angler baitfishing with light tackle. At low tide, gather your natural bait: piling worms, mussels, shore crabs, snails and clams—fun for the whole family. Later impale these baits on small hooks, and fish near the bottom. Conservation-minded anglers release their large, fat, pregnant striped seaperch alive to ensure future catches.

 Divers often notice schools of striped seaperch in shallow areas around weed-covered rocks or near pilings and jetties. During summer the female releases her young, conceived the previous year. The newly born, each between 4 and 6 cm long, huddle in shallow bays, while every adult female awaits the ritualistic advances of an attentive male.

 Marketed with various other varieties of seaperch, this species does constitute a minor portion of the California catch but is of no commercial value in the Pacific Northwest where its shallow, rocky habitat renders the striped seaperch inaccessible to net fishermen.

 Even from a wharf, above water, this striped and darker seaperch is readily distinguishable from other silvery, lighter varieties. Watch the powerful pectoral strokes propel the deep body while the fish slowly and deliberately grazes small animals from rocky outcroppings, particularly in summer.

 Fair eating but loose hair-like bones as well as the vertebral column require attention. Fillet and chill before cooking; use in any recipe calling for mild-tasting fish.

40 Pile Perch

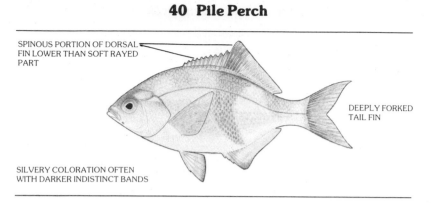

SPINOUS PORTION OF DORSAL FIN LOWER THAN SOFT RAYED PART

DEEPLY FORKED TAIL FIN

SILVERY COLORATION OFTEN WITH DARKER INDISTINCT BANDS

SPECIES: *Rhacochilus vacca*—from the Greek *racos* and *cheilus*, meaning "ragged" and "lip;" and the Latin *vacca*, "cow."

ALTERNATE NAMES: pile seaperch, pile surfperch, dusky sea-perch, dusky perch, forktail perch, splittail perch, silver perch*, porgy*.

MAXIMUM RECORDED SIZE: 44.2 cm (17.4 inches).

DISTRIBUTION: Isla San Martin, northern Baja California, Mexico, to Wrangell, southeastern Alaska.

A pile perch fishing trip is fine family sport, particularly in summer. First go to the shore at low tide and gather shore crabs, mussels, snails, piling worms, or clams for bait. Later, as the tide comes in, fish for the pile perch from a wharf or jetty along a shallow bay or shoreline. Surprisingly strong, the light-weight pile perch fights hard when hooked, particularly on light tackle. A lesson in conservation, finally, should occur when someone in the party catches and releases a large pregnant female. Any specimens caught may be kept alive and used as bait for lingcod or rockfish later.

Particularly during summer dives near pilings, over shallow reefs, or along shorelines and at depths usually less than 20 m (66 feet), neoprene-suited explorers frequently encounter the silvery, schooling pile perch. Notice the ritualistic courtship displays as the aggressive, darker males roll over on their sides and flash in front of the passive, more lightly-hued females.

Commonly filleted and sold fresh as "perch," the pile perch forms a very minor, somewhat incidental catch for Pacific Northwest trawlers and is of only slightly greater interest to Californian harvesters.

Readily recognized, even from above water, as the large, silvery, deep-bodied fish with broad irregular dark bands on its sides, the pile perch commonly gathers around piers and pilings, but is seldom trapped in tidepools.

Many small bones notwithstanding, the fine-grained flesh of the pile perch provides fair eating.

*—incorrect

41 White Seaperch

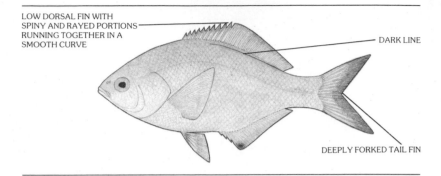

LOW DORSAL FIN WITH
SPINY AND RAYED PORTIONS
RUNNING TOGETHER IN A
SMOOTH CURVE

DARK LINE

DEEPLY FORKED TAIL FIN

SPECIES: *Phanerodon furcatus*—from the Greek *phaneros* and *odons*, meaning "evident" and "teeth;" and the Latin *furcatus*, "forked."

ALTERNATE NAMES: white perch, silver seaperch*.

MAXIMUM RECORDED SIZE: 32 cm (12.5 inches).

DISTRIBUTION: Punta Cabras, northern Baja California, Mexico, to southern Vancouver Island, southern British Columbia.

Wharves and jetties extending over sandy substrates are good angling platforms to bottomfish for the white seaperch, especially during summer. Begin a successful fishing session by gathering its natural intertidal prey for bait; clams, snails, piling worms or small shrimps. Once hooked on light line, the energetic white seaperch provides good sport and itself is good bait for large bottomfish.

Search along sandy bottoms of shallow bays when seeking the white seaperch and pay particular attention to pilings and other shelter. Upon close and careful approach to a specimen, study the low dorsal fin contour which distinguishes this species from the more common, usually darker, pile perch. When night falls, the white seaperch huddles close to the bottom and the diver can then stalk it more easily for specific identification.

Not really abundant in the Pacific Northwest, the small white seaperch is mainly harvested in California where it is netted in shallow water and sold with other surfperch under the catch-all term "perch."

Although large tidepools might occasionally trap the white seaperch, it is more often around floats or jetties. Specific identification from above water, though, proves difficult because this species resembles very much the common, darker pile perch. Be patient though, for quite frequently when they are feeding or breeding, both species tilt on their sides and present a better view.

Panfrying the batter-coated, fine-grained flesh of the white seaperch enhances its delicate flavour. Be wary of small bones, even in fillets.

*—incorrect

42 Silver Surfperch

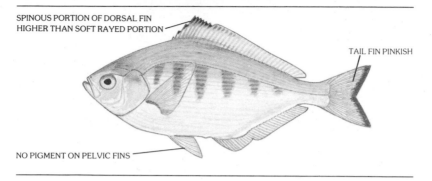

SPINOUS PORTION OF DORSAL FIN
HIGHER THAN SOFT RAYED PORTION

TAIL FIN PINKISH

NO PIGMENT ON PELVIC FINS

SPECIES: *Hyperprosopon ellipticum*—from the Greek *hyper* and *prosopon*, meaning "above" and "face;" and the Latin *ellipticum*, "elliptical."

ALTERNATE NAMES: silver perch, porgy *.

MAXIMUM RECORDED SIZE: 27 cm (10.5 inches).

DISTRIBUTION: Rio San Vincente, northern Baja California, Mexico, to Schooner Cove, near Ucluelet, Vancouver Island, southern British Columbia.

Not often specifically sought, the small silver surfperch occasionally shows up in the creels of shore bound anglers fishing surf-swept, open-coast beaches or adjacent bays. Summer finds large pregnant silver surfperch congregating in shallow, calm, weed-choked bays and giving birth to as many as 30 tiny, 2-cm young. By releasing any of these swollen mothers-in-waiting, the angler helps to ensure future supplies of this silvery, schooling fish. A thoughtful angler might retain other specimens in a bucket of seawater, then use them as live bait for larger quarry.

The quick, elusive silver surfperch resides along exposed, surf-swept beaches or nearby shallow coves, and is potentially, at least, directly in the diver's path. However, turbid summer conditions permit only fleeting glimpses of this schooling fish, no matter how carefully stalked. A night sortie during which a diver can illuminate and virtually "freeze" the silver surfperch in a beam of light offers a slightly better viewing opportunity.

So rarely do commercial fisherman set nets in the habitat of the unmarketable silver surfperch that it is seldom so much as an incidental catch.

Although the silver surfperch swims in shallow water around open coast beaches and jetties, do not expect to see it: the turbid water obscures it.

Dlll

The culinary value of the bone-laden silver surfperch is somewhat reduced by its small size: only thin, short fillets slice from even the largest fish.

* —incorrect

63

43 Redtail Surfperch

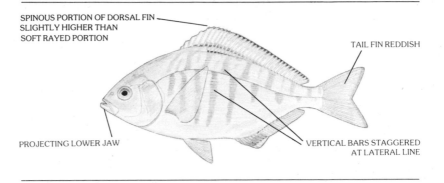

SPINOUS PORTION OF DORSAL FIN
SLIGHTLY HIGHER THAN
SOFT RAYED PORTION

TAIL FIN REDDISH

PROJECTING LOWER JAW

VERTICAL BARS STAGGERED
AT LATERAL LINE

SPECIES: *Amphistichus rhodoterus*—from the Greek *amphi*, *stoichus* and *rhodoterus*, meaning "double," "series" and "rosy."

ALTERNATE NAMES: redtail surf perch, redtail seaperch, redtail perch, redtail, porgy*.

MAXIMUM RECORDED SIZE: 41 cm (16 inches).

DISTRIBUTION: Avila Beach, central California, to Hesquiat Bay, central Vancouver Island, British Columbia.

Surfcast for schooling redtail surfperch along sandy beaches directly exposed to the Pacific surge, or from adjacent piers or jetties and use the natural dietary items of this fish for bait: razor clams, ghost shrimps, sand shrimps, mussels, shrimps, shore crabs and sand crabs. Or try hooking a piling worm on a small spinner, casting, and slowly retrieving it. Particularly when caught on light tackle, the redtail surfperch fights tenaciously. This species is an excellent bait for larger fish, but as a good conservation practice, do release any obviously pregnant females.

Unfortunately the open, coastal, surf-swept beaches create a surging hardship for the diver but are prime redtail surfperch habitat; shallow adjacent bays, however turbid, offer moderately better viewing conditions. By day this active, schooling species is difficult to approach, but at night with the aid of the beam from an underwater flashlight, the diver may move a little closer.

Trawled or gillnetted along shallow beaches, and sold fresh with other seaperch as "perch," the redtail surfperch forms only a minor part of the annual Californian catch. Not often caught or marketed in the Pacific Northwest.

Occasionally if the water is clear, the jettyside naturalist may see the strong-swimming redtail surfperch propelling itself along with powerful pectoral fin strokes, but there is usually some difficulty distinguishing this fish from other silvery surfperch.

A tasty but only fair pan fish. Fillet it to remove those many fine, hair-like bones.

*—incorrect

44 Kelp Perch

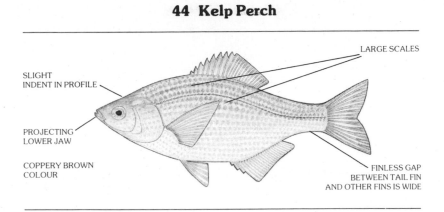

SLIGHT
INDENT IN PROFILE

LARGE SCALES

PROJECTING
LOWER JAW

COPPERY BROWN
COLOUR

FINLESS GAP
BETWEEN TAIL FIN
AND OTHER FINS IS WIDE

SPECIES: *Brachyistius frenatus*—from the Greek *brachys* and *istion*, meaning "short" and "sail;" and the Latin *frenatus*, "bridled."

ALTERNATE NAMES: kelp sea-perch, kelp surfperch, brown sea-perch, brown perch.

MAXIMUM RECORDED SIZE: 22 cm (8.5 inches).

DISTRIBUTION: Guadalupe Island, off central Baja California, Mexico, to Welcome Harbour, Porcher Island, northern British Columbia.

 Seldom sought or caught, the small kelp perch commonly feeds upon minute worms, shrimps, clams, crabs, or other shrimp-like creatures that share its dense kelp bed habitat. If obtaining some live kelp perch, try using them as bait for large bottomfish such as rockfishes or greenlings.

 When cruising along shallow shorelines with large, thick growths of kelp, look very closely among the fronds to find the golden brown kelp perch at depths less than 27 m (90 feet). Either solitary or as a member of a small group, each one nibbles at the many tiny animals living on and among the lush golden brown plants. Also known to be a "cleaner fish," the bold kelp perch may fastidiously pick external parasites from the bodies of larger fishes; these "clients" even obligingly erect their fins and remain motionless while being relieved of their annoyances.

 Although the small, unmarketable kelp perch is effectively isolated from most commercial fishing activities by its shallow kelp-choked habitat, kelp harvesting machines incidentally capture considerable numbers, especially in California.

 The kelp perch often swims in clear water just below the surface, but darts quickly among the kelps where it is difficult to see. Rarely do adult kelp perch or the summer-born juveniles become trapped in tidepools.

 Even the largest kelp perch provides the chef with little usable flesh.

45 Shiner Perch

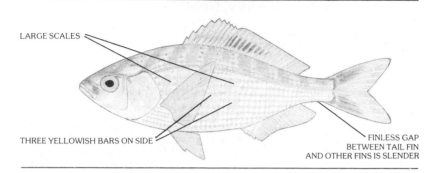

LARGE SCALES

THREE YELLOWISH BARS ON SIDE

FINLESS GAP
BETWEEN TAIL FIN
AND OTHER FINS IS SLENDER

SPECIES: ***Cymatogaster aggregata***—from the Greek *cymos* and *gaster*, meaning "foetus" and "belly;" and the Latin *aggregata*, "crowded together."

ALTERNATE NAMES: shiner seaperch, shiner surfperch, seven eleven perch, yellow shiner, shiner, seven eleven, pogy.

MAXIMUM RECORDED SIZE: 18 cm (7 inches).

DISTRIBUTION: Bahia San Quintin, northern Baja California, Mexico, to Wrangell, southeastern Alaska.

A very popular summer quarry for children fishing from wharves, the easily caught shiner perch swarms in shallow, sandy-bottomed bays. Attach marine worms, mussels, snails, shrimps, or clams to tiny hooks, look for a school of shiner perch swimming just below the surface, and you may easily fill a bucket with them. Use these excellent live baits for larger and more sporty targets.

Explore shallow, weedy bays or around pilings during spring and summer to find the abundant schooling shiner perch while it engages in breeding behaviour. After the large females have each released as many as 17 fully-developed young, the very aggressive, temporarily black males, some themselves just newly born, swim rapidly in a bouncing fashion around the females to attract their attention. Ultimately approaching from below and behind, these eager males copulate with the females and assure next year's brood. At night look for shiner perch in the beam of your underwater flashlight as they huddle against the sandy bottom.

While not actively sought by commercial fishermen, the common shiner perch nonetheless shows up abundantly in shrimp trawl nets towed at depths to 146 m (480 feet).

In summer, schools of shiner perch swarm in shallow, weedy areas, particularly around pilings, under floats or even in large tidepools.

Although most people do not bother with the small shiner perch, oriental North Americans eat it dried or pickled.

THE GOBIES
(Family: Gobiidae)

One of the largest fish families, the very successful gobies or Gobiidae, number in excess of seven hundred known living species. While most live in shallow to moderately deep coastal marine habitats, a few exceptional species live in brackish or even freshwater haunts. Sandy, silty bays and tidal flats are prime territory for most gobiids, but numerous others colonize rocky, more solid substrates. Most gobies thrive in the warm waters of tropical latitudes, particularly in the Indo-Pacific, the Mediterranean and Caribbean; only a few tolerate temperate climates.

Active bottom-dwelling, or benthic, creatures, the abundant gobies, few of which attain a length greater than 15 cm (6 inches), swim in short bursts punctuated by frequent—often brief—rest stops. They rest on the bottom, their length supported by their pectoral fins and by their distinctive cone-shaped pelvic fins at the front and the tail fin at the back. Certain gobiids actively burrow in the sand or mud, creating tunnels to shelter themselves from predators, while other resourceful gobiids seek out the existing burrows of worms, shrimps and crabs, often establishing very special relationships with these invertebrate "tunnel mates."

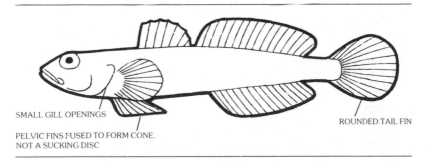

SMALL GILL OPENINGS

PELVIC FINS FUSED TO FORM CONE.
NOT A SUCKING DISC

ROUNDED TAIL FIN

Remains of ancient gobies, fossilized in prehistoric sedimentary rock, date the group at least as far back as the Eocene epoch, 25 million years ago. Very distinctive animals, the gobiids possess no very close living relatives. Generally people eat only the large gobies, such as the gargantuan sleeper of 61 cm (24 inches), but the Russians use an unusually small goby as a preserve in tomato sauce—called *bichki*—and Philippine natives make the minute larvae of others into a fish paste. Incidentally, the world's smallest fish is a tiny gobiid which lives in the Philippines and and is no longer than 11 mm!

46 Blackeye Goby

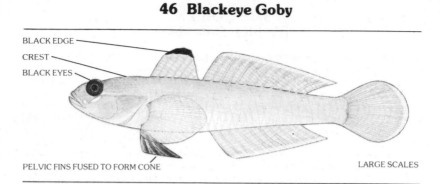

BLACK EDGE
CREST
BLACK EYES

PELVIC FINS FUSED TO FORM CONE

LARGE SCALES

SPECIES: *Coryphopterus nicholsi*—from the Greek *coryphos* and *pteros*, meaning "head" and "fin;" and honouring its discoverer, Captain Henry E. Nichols of HM *Hassler*.

ALTERNATE NAMES: crested goby, large-scaled goby, bluespot goby.

MAXIMUM RECORDED SIZE: 15 cm (6 inches).

DISTRIBUTION: Punta Rompiente, central Baja California, Mexico, to Wales Island, northern British Columbia.

Very occasionally while angling from a wharf or jetty, a bottom-fisherman using a tiny baited hook might unintentionally catch the blackeye goby.

By flaring its gill covers and raising its bright fins, the aggressive blackeye goby often boldly challenges the diver. Look for this common species where soft bottoms grade into loose rocky substrates or reefs. During the April to October breeding season, the mature and even more aggressive male claims a crevice, often removing any excess sand or silt. Later he lures any ripe female into this den by erecting all his fins, and repeatedly rising just off the bottom to pause in midwater briefly before settling back to the sea floor. Once inside, the female deposits up to 1,700 minute pink eggs. The male tenaciously guards them for several weeks until they hatch and release tiny transparent larvae which will reside temporarily at the ocean's surface.

The unsaleable blackeye goby sometimes ventures into prawn traps, or finds itself scooped up in fine-meshed shrimp trawl nets towed at depths less than 106 m (348 feet).

A very infrequent tidepool hostage, the bright orange to pale pink blackeye goby often lives in very shallow water, where some rock is present. Dockside observers may also view this short, robust fish resting on the bottom near natural or man-made cover. A predator of snails and tiny shrimps, the blackeye goby itself falls prey to various rockfishes.

Too small for most recipes, the unappetizing blackeye goby is nevertheless edible.

47 Bay Goby

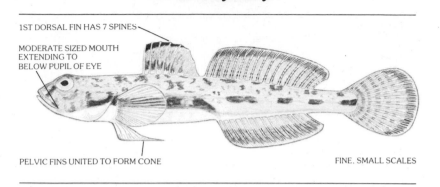

1ST DORSAL FIN HAS 7 SPINES

MODERATE SIZED MOUTH
EXTENDING TO
BELOW PUPIL OF EYE

PELVIC FINS UNITED TO FORM CONE

FINE. SMALL SCALES

SPECIES: *Lepidogobius lepidus*—from the Greek *lepis* and *gobio*, meaning "scale" and referring to another small fish; and the Latin *lepidus*, "pretty."

ALTERNATE NAMES: finescale goby, fine-scale goby.

MAXIMUM RECORDED SIZE: 10 cm (4 inches).

DISTRIBUTION: Isla Cedros, central Baja California, Mexico, to Welcome Harbour, northern British Columbia.

Very seldom does an angler catch the small, tiny-mouthed bay goby; if available, though, it makes excellent bait for flounders, codfishes or greenlings.

Almost transparent, the light-gray bay goby is seldom obvious to even the few divers who explore its typical muddy or silty, level bottom habitat. Even if noticed this quick little creature warily swims across the easily stirred-up bottom then often retreats down a tiny hole in the silt, just before the diver gets a good look. Actually, the resourceful bay goby often shares a burrow with a worm, a geoduck, or a mud shrimp that has excavated the hole.

Shrimp trawlers who drag over shallow, soft bottoms only rarely capture the bay goby because, upon noticing the approaching trawl, this crafty little fish retreats safely down a nearby burrow. In many locales at depths to 210 m (660 feet), the bay goby is so common that a swarm may be recorded on echo sounding equipment.

Observant beach strollers sometimes notice the bay goby resting on the sand or among the sea lettuce or eelgrass growing in tidal flats and estuaries. Watch carefully, though, because a startled specimen may dart quickly down a nearby burrow to escape, not distinguishing a beachcomber from one of its natural predators, such as the marauding staghorn sculpin or a shore bird. The shallow-dwelling bay goby also lives on silty bottoms under floats or around pilings.

Try a Russian idea: pickle whole bay gobies, then serve them with tomato sauce.

48 Arrow Goby

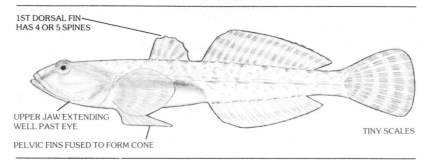

1ST DORSAL FIN HAS 4 OR 5 SPINES

UPPER JAW EXTENDING WELL PAST EYE

TINY SCALES

PELVIC FINS FUSED TO FORM CONE

SPECIES: *Clevelandia ios*—honouring Dr. Daniel Cleveland, a past president of the San Diego Society of Natural History and the Latin *ios*, meaning "arrow."

ALTERNATE NAMES: none.

MAXIMUM RECORDED SIZE: 6.4 cm (2.5 inches)

DISTRIBUTION: San Bartholomewe Bay, in the Gulf of California, around Baja California, Mexico, north to Rivers Inlet, central British Columbia.

Readily accessible at low tide and abundant in many areas, the tiny arrow goby is an excellent live bait for the bottom-fishing angler.

The light gray, almost transparent arrow goby flourishes along gently sloping muddy tidal flats and adjacent tidepools or lagoons where very few divers prowl.

Because few commercial fishermen harvest along silty tidal flats where the minute arrow goby dwells, this unexploited animal is never even an unintentional catch.

Look closely along shallow mud flats and see tiny ghost-like arrow gobies dart away before disappearing down nearby clam, shrimp or worm burrows. An interesting, mutually beneficial relationship, known as symbiosis, exists between this active fish and these creatures that excavate the burrows; for example, the slender arrow goby consumes scraps of food left by the feeding shrimp and thereby keeps the shared "lodging" clean. In times of danger when kingfishers, terns, staghorn sculpins or whitespotted greenlings are on the prowl, the wary arrow goby may bury itself using rapid movements of its fins, gill covers and body. Although the bright summer sun may warm its shallow habitat and heavy winter rains often dilute the surrounding salt water, this hardy creature may possibly live a two- or three-year life span.

Perhaps the only possible culinary use for the tiny arrow goby is as an ingredient in fish broth or paste.

THE RONQUILS
(Family: Bathymasteridae)

The ronquils and the searchers comprise the tiny family known scientifically as the Bathymasteridae, which translates to "deep searcher family." At present a subject of review, the poorly studied group contains only about seven living species. Living only within the cool, temperate waters of the North Pacific, the small ronquils never grow longer than 30 cm (12 inches). Marine fishes all, bathymasterids live at shallow to moderate depths along coastal regions and often establish dens where rocky outcroppings meet sandy or muddy substrates.

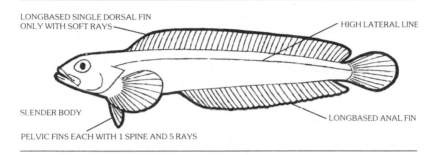

LONGBASED SINGLE DORSAL FIN ONLY WITH SOFT RAYS

HIGH LATERAL LINE

SLENDER BODY

LONGBASED ANAL FIN

PELVIC FINS EACH WITH 1 SPINE AND 5 RAYS

As yet, paleontologists have not found any fossils definitely known to be prehistoric remains of ronquils and scientists can only speculate about bathymasterid origins. Interrelationships between ronquils and other present-day fishes are debatable, but some ichthyologists think that the tropical jawfishes rate as the bathymasterids' closest living kin. Too small and insignificant for commercial exploitation, the little-known ronquils do serve as forage species for many large, economically important groundfishes in the North Pacific.

Although collections indicate that the smooth ronquil, with fewer unbranched soft dorsal fin rays than the northern ronquil, page 72, exists in the Pacific Northwest, it is apparently not common there; a further complication is the possible existence of a third very similar species. Another, the searcher, is not included here because it has a southern distribution limit of the Queen Charlotte Islands, northern British Columbia, as does the bluefin searcher.

49 Northern Ronquil

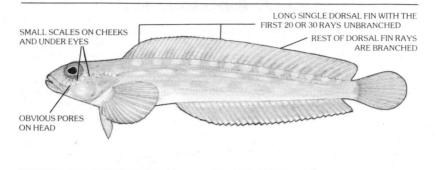

SMALL SCALES ON CHEEKS
AND UNDER EYES

LONG SINGLE DORSAL FIN WITH THE
FIRST 20 OR 30 RAYS UNBRANCHED

REST OF DORSAL FIN RAYS
ARE BRANCHED

OBVIOUS PORES
ON HEAD

SPECIES: *Ronquilus jordani*—from the Spanish *ronquilus*, a traditional name for a similar fish; and honouring David Starr Jordan, considered the father of American ichthyology.

ALTERNATE NAMES: ronquil.

MAXIMUM RECORDED SIZE: 18 cm (7 inches).

DISTRIBUTION: Monterey Bay, central California, to the Aleutian Islands and the Bering Sea, Alaska.

Although unknown by most anglers, the usually unsought northern ronquil may occasionally seize small hooks baited with tiny pieces of fish or small invertebrate animals. Such an unusual incidental catch will likely occur from a float or pier located over muddy, sandy bottoms. Try northern ronquil as a bait for larger quarry.

Search along silty or sandy substrates near adjacent rock formations, flat rocky bottoms, or other shelter including cans, jars, and rubber tires, for the active northern ronquil. This wary creature usually retreats into its secure lair and pokes its head out again to watch you. During February and March the anal fin of the mature male northern ronquil becomes bright blue and yellow and his head develops some dark purple blotches. Enticed by his bright coloration and his long intricate courtship displays, a co-operative gravid female deposits adhesive amber-coloured eggs, usually upon a flat loose rock where the male fertilizes them. Within several weeks, the transparent, sliver-like larvae hatch, and then forage near the surface for almost a month.

Too small to be of direct economic value, the common northern ronquil may incidentally enter commercial catches either as captives in fine-meshed shrimp trawl nets or as hostages in prawn traps set at depths to 180 m (694 feet).

Only seldom does the dockside naturalist spy the elongate, pale northern ronquil. It is almost never trapped in tidepools.

Although edible, the northern ronquil is not recommended because its size makes it not worth the effort.

THE KELPFISHES
(Family: Clinidae)

The Clinidae, which translates to "blenny family," is a very large group containing more than one hundred and seventy-five recognized living species of kelpfishes, klipfishes, pikeblennies, fringeheads and scaleless blennies. Although cool temperate waters of both the northern and southern hemispheres shelter an assortment of these small, secretive animals, most of them abound in warm tropical waters, particularly the Indo-Pacific. The various and variable clinids live strictly in the marine environment; most species are shallow, coastal denizens that constantly huddle among rocks encrusted with colourful seaweeds and animals.

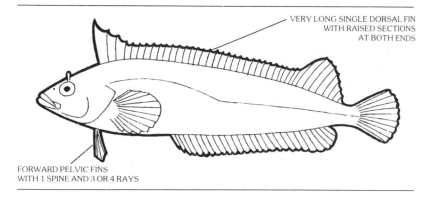

VERY LONG SINGLE DORSAL FIN
WITH RAISED SECTIONS
AT BOTH ENDS

FORWARD PELVIC FINS
WITH 1 SPINE AND 3 OR 4 RAYS

Masters of camouflage, the attractively marked clinids rapidly alter their colour, both pattern and intensity, in order to obscure their body shape and to blend exactly with their underwater surroundings. They change very quickly when wandering from one distinctly coloured or patterned background to another. Not only does their alterable camouflage conceal them from numerous hungry predators, but it is also a definite advantage to these clinids lurking motionless in wait for their own prey.

The ancestors of present-day kelpfishes, according to dated clinid fossils, flourished in habitats similar to those of modern species, nearly 30 million years ago in the Eocene epoch. Together with the gunnels, pricklebacks, wolffishes, wrymouths, quillfishes, and graveldivers—six other families of long slender fishes—the kelpfishes are members of a large group known as the blennies, all having single, very long dorsal fins. Because the small kelpfishes—only one species, the giant kelpfish of California, attains a length of 61 cm (24 inches)—inhabit very shallow and generally unfished waters, they are commercially insignificant. Only a few scientists and aquarists value them.

50 Striped Kelpfish

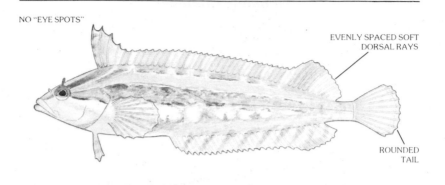

NO "EYE SPOTS"

EVENLY SPACED SOFT
DORSAL RAYS

ROUNDED
TAIL

SPECIES: *Gibbonsia metzi*—honouring William P. Gibbons, an early naturalist from Alameda, California, and Charles W. Metz, a student of ichthyology once fascinated by kelpfishes.

ALTERNATE NAMES: striped kelp-fish.

MAXIMUM RECORDED SIZE: 24 cm (9.5 inches).

DISTRIBUTION: Hondo Canyon, near Ensenada, northern Baja California, Mexico, to Maquinna Point, Nootka Sound, Vancouver Island, British Columbia.

The tiny-mouthed striped kelpfish is a most unusual and unsought catch for baitfishermen because few such anglers brave the Pacific surge as it crashes on weed-covered and slippery rocks. This fish shows little potential as live bait because of its secretive habits and dark coloration.

An adventurous diver daring to explore surf-swept rocky shallows of the exposed Pacific shore might easily drift over many well-camouflaged striped kelpfish without even noticing them along weed-choked rocky crevices. On a calm day carefully search large tidepools and adjacent waters at depths to 9 m (30 feet) and perhaps find a specimen amid the colourful swaying red, brown, and green plantlife.

Commercial fishermen never set gear in the striped kelpfish's shallow or intertidal realm because no marketable species thrive there.

Search carefully along the weedy shallows, or in tidepools of the exposed Pacific coast and perhaps notice the secretive, seldom-seen striped kelpfish. To catch them, try thrusting a long-handled, fine-meshed dipnet through large kelps and other plants where young specimens most often reside. Hatched in spring, striped kelpfish larvae live in surface waters before becoming miniature adults and settling out into their permanent weedy, rocky bottom habitat. The female usually outgrows the male.

No one has ever eaten a striped kelpfish—probably.

51 Crevice Kelpfish

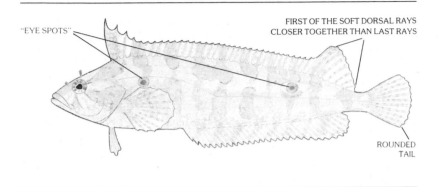

"EYE SPOTS"

FIRST OF THE SOFT DORSAL RAYS
CLOSER TOGETHER THAN LAST RAYS

ROUNDED
TAIL

SPECIES: *Gibbonsia montereyensis*—honouring William P. Gibbons, an early naturalist from Alameda, California, and the location for an early collection of this fish.

ALTERNATE NAMES: spotted kelpfish*.

MAXIMUM RECORDED SIZE: 11 cm (4.5 inches).

DISTRIBUTION: Bahia San Tomas, northern Baja California, Mexico, to Ucluelet, Vancouver Island, southern British Columbia.

The small crevice kelpfish, with its minute mouth, is almost an impossible catch for the bait-fisherman because it lives in shallow, weedy, rocky shores along open coasts where few anglers dare to venture. It makes only fair bait itself.

Theoretically, divers could encounter the shallow-dwelling crevice kelpfish from the intertidal zone to a depth of 21 m (70 feet); sightings very rarely occur, though, because this tiny, often stationary creature changes its colour to blend with any background it rests upon. Swirling sand, plant debris, and plankton blooms cause poor visibility in the crevice kelpfish's rugged realm, further inhibiting observation.

Too small for the marketplace, the crevice kelpfish lives in an environment commercial fishermen never harvest in; consequently, it is never caught.

To find the secretive crevice kelpfish, take a strong, fine-meshed dipnet with you along weedy, rocky shores directly exposed to oceanic waves. If a number of keen, active beachcombers are present at low tide, try bailing out a tidepool before the tide comes back in, and perhaps find a crevice kelpfish or two. After watching and studying the various fishes and other creatures, return them carefully so others might enjoy them too; the returning water will fill the empty tidepool and rejuvenate the temporarily exposed animals.

Its small size and rare capture keep the crevice kelpfish off the stove.

*—incorrect

THE PRICKLEBACKS
(Family: Stichaeidae)

The pricklebacks, cockscombs, warbonnets, shannies and eelblennies total fifty-four recognizable living species and form the small family scientifically termed the Stichaeidae, translating to "blenny family." All stichaeids are marine creatures living in cool or frigid coastal waters of the northern hemisphere. The vast North Pacific Ocean is where most of these live, but a few species flourish in the Arctic and North Atlantic. While all live on the bottom, some pricklebacks reside in shallow, intertidal habitats, and still others colonize rocky, weedy cliff faces or gently sloping, muddy sea floors; few prefer deep, silty locales.

Prehistoric pricklebacks first existed in the seas of this planet during the Eocene epoch, nearly 30 million years ago, according to available stichaeid fossils known at present. Today though, other elongate northern blenny fish families, such as the gunnels, wrymouths and wolffishes, are the modern stichaeids' closest living relatives; all have long dorsal and anal fins which contain at least some spines. While the small, slender pricklebacks—few growing to 51 cm (20 inches)—are not directly important to commercial fishing interests, they do contribute, through the food chain, to important fisheries.

LONG DORSAL FIN WITH SHARP SPIKE-LIKE SPINES (PRODUCING "PRICKLE" BACK)

LATERAL LINE: COMPLETE.
PARTIAL, ABSENT OR 4

PELVIC FINS (IF PRESENT)
WITH 1 SPINE AND 3 OR 4 RAYS

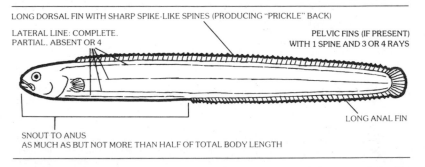

LONG ANAL FIN

SNOUT TO ANUS
AS MUCH AS BUT NOT MORE THAN HALF OF TOTAL BODY LENGTH

Several other pricklebacks are taken, although rarely, in the Pacific Northwest. One lives at inaccessible depths: the bluebarred prickleback, with numerous long white bars along its sides and black spots on the rear of its dorsal fin. Three others dwell in reasonably shallow waters but are, for inexplicable reasons, hardly ever seen: the Y-prickleback, a small species with Y-shaped marks along its sides and spots along its dorsal fin; the pearly prickleback, a newly discovered species with a complex bushy cirrus between its eyes; and the daubed shanny, similar to the fish on page 82, but with the lower rays of the pectoral fins being web-less and bare. Two other pricklebacks, the matcheek warbonnet and the Arctic shanny live primarily in northerly regions.

52 Mosshead Warbonnet

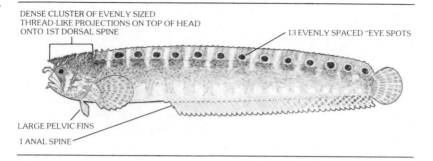

DENSE CLUSTER OF EVENLY SIZED
THREAD-LIKE PROJECTIONS ON TOP OF HEAD
ONTO 1ST DORSAL SPINE

13 EVENLY SPACED "EYE SPOTS"

LARGE PELVIC FINS

1 ANAL SPINE

SPECIES: *Chirolophis nugator*—from the Greek *chiros* and *lophis*; meaning "hand" and "crest;" and the Latin *nugator*, "elegant" or "fop."

ALTERNATE NAMES: mosshead prickleback, ornamented blenny.

MAXIMUM RECORDED SIZE: 15 cm (6 inches).

DISTRIBUTION: San Miguel Island, southern California, to Umnak Island, in the Aleutian chain, Alaska.

Although potentially very useful as a live bait for large bottomfish, the tiny, never-caught mosshead warbonnet is unknown to and unattainable by most anglers.

A popular species with underwater photographers, the attractive mosshead warbonnet lives in various shallow-water locales, but especially inside empty shells or holes along rocky shores at depths less than 20 m (66 feet). If diving around piers or jetties, pay particular attention to the interiors of discarded bottles, cans or pipes, for often only the fluffy-topped head of this creature pokes out. Once disturbed, the skittish mosshead warbonnet streaks off to more secure surroundings nearby.

A rare incidental catch in fine-meshed prawn traps set in shallow locales near rocky areas, the tiny mosshead warbonnet is of no direct economic value.

At the very lowest of tides, the curious beachcomber may occasionally find dark mosshead warbonnets living among crevices in rocky tidepools. Because this fish is secretive and difficult to see, the best way to study it is by first bailing out the water, exposing the trapped specimens, and then placing them in a bucket of that same seawater. After enjoying a close encounter, return any captives to nearby full pools, or to the adjacent water's edge. Even if the water is clear, dockside observers rarely sight the well-camouflaged mosshead warbonnet hiding in empty barnacle casings or other shelter attached to pilings or the undersides of floats.

The small mosshead warbonnet is not a practical item for the menu.

53 Decorated Warbonnet

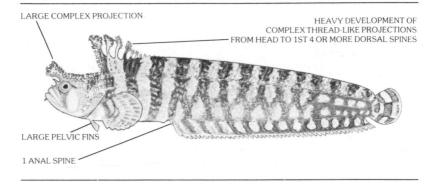

LARGE COMPLEX PROJECTION

HEAVY DEVELOPMENT OF
COMPLEX THREAD-LIKE PROJECTIONS
FROM HEAD TO 1ST 4 OR MORE DORSAL SPINES

LARGE PELVIC FINS

1 ANAL SPINE

SPECIES: *Chirolophis decoratus*—from the Greek *chiros* and *lophis*, meaning "hand" and "crest;" plus the Latin *decoratus*, "ornamental."

ALTERNATE NAMES: decorated prickleback.

MAXIMUM RECORDED SIZE: 42 cm (16.5 inches).

DISTRIBUTION: Humboldt Bay, northern California, to the Bering Sea and Bering Strait coasts of Alaska, along the Aleutian chain to Kamchatka, USSR.

Although almost never caught and therefore unknown to most anglers, the elongate decorated warbonnet would be good live bait for rockfishes and lingcod.

An often-photographed species living at depths between 15 and 91 m (50 and 300 feet), the wary decorated warbonnet usually looks out at the diver from among rocky crevices, or the openings of large, hollow sponges. Photographers particularly enjoy framing its distinctively bushy head against the dark, hollow recesses of its lair. Stalk it carefully, though, for once startled the camera-shy decorated warbonnet will often retreat deeper inside its shelter or bolt away to another sheltering site; do not try to follow this reluctant subject but rather wander along and hope to find another undisturbed model. Shrimps and their kin, with which the lurking decorator warbonnet frequently shares its cave or sponge, comprise the bulk of the diet of this brownish fish.

Professional prawn fishermen who set their traps at appropriate depths often find the slippery, writhing decorated warbonnet as an incidental, unmarketable catch. While most commercial fishermen discard or ignore it, other curious and thoughtful ones donate this fascinating, easily kept species to public aquariums, where it makes a wonderful display. Just place the specimen in a bucket of seawater, refreshing it periodically before contacting grateful aquarium officials.

The subtidal decorated warbonnet inhabits depths inaccessible to the dockside observer or beachcombing naturalist.

Very few seafood lovers have the opportunity to try the eel-shaped decorated warbonnet's undoubtedly palatable flesh.

54 High Cockscomb

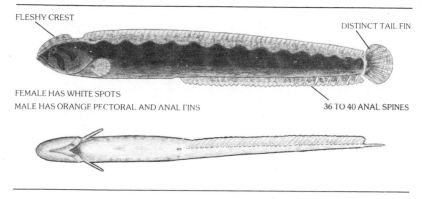

FLESHY CREST

DISTINCT TAIL FIN

FEMALE HAS WHITE SPOTS
MALE HAS ORANGE PECTORAL AND ANAL FINS

36 TO 40 ANAL SPINES

SPECIES: *Anoplarchus purpurescens*—from the Greek *anoplos* and *archos,* meaning "unarmed" and "anus;" plus the Latin *purpureus,* "purple."

ALTERNATE NAMES: cockscomb prickleback, crested blenny, cockscomb.

MAXIMUM RECORDED SIZE: 20 cm (7.8 inches).

DISTRIBUTION: Santa Rosa Island and Trinidad Bay, southern California, to Attu Island in the Aleutian chain, and the Pribilof Islands, Bering Sea.

Readily gathered in quantity and kept alive in a bucket of frequently-changed seawater, the active high cockscomb makes truly excellent bait when bottom-fishing for large rockfish or greenlings.

A very common inhabitant of water less than 3 m (10 feet), along rocky shorelines or in tidepools, the shy high cockscomb usually huddles under rocks, and only rarely is exposed completely. If seen at all, only its head may be visible as it peers from beneath a boulder or from inside a bottle. If this shelter is overturned, the variably coloured high cockscomb rapidly scurries away and wriggles underneath another shelter. This secretive fish apparently spends some time in the open at night; be particularly vigilant on a nocturnal dive.

Of no direct economic value, the small high cockscomb is not caught because commercial fishermen never set their gear in its intertidal or shallow, rocky habitat.

Shore-strolling naturalists overturning rocks intertidally nearly always find the writhing, high cockscomb in great abundance. To grasp or catch a specimen, however, is a frustrating, slippery experience; all too often the flopping creature slithers quickly out of sight. Many predators, including one variety of garter snake forage for the plentiful high cockscomb at low tide. During late winter, a female high cockscomb may coil around her cluster of whitish eggs and fan them with her caudal and pectoral fins until they hatch.

Too small to be useful for the practical chef.

55 Slender Cockscomb

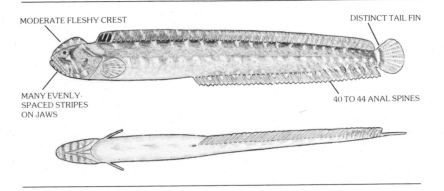

MODERATE FLESHY CREST

DISTINCT TAIL FIN

MANY EVENLY-
SPACED STRIPES
ON JAWS

40 TO 44 ANAL SPINES

SPECIES: *Anoplarchus insignis*—from the Greek *anoplos* and *archos*, meaning "unarmed" and "anus;" plus the Latin *insignis*, "no signal."

ALTERNATE NAMES: cockscomb.

MAXIMUM RECORDED SIZE: 12 cm (4.8 inches).

DISTRIBUTION: Arena Cove, Mendocino County, northern California, to Attu Island, in the Aleutian chain, Alaska.

The seldom-seen slender cockscomb would be a good live bait if it were available, but it lives in deep water and anglers never take it.

While the secretive slender cockscomb's rocky habitat is from the surface to a diveable 30 m (100 feet), even an observant aquanaut seldom knowingly confronts this species because it rarely leaves the security of its rocky crevice. Even if diligently looking under movable rocks, the diver may have only a fleeting glimpse of this elusive creature because it quickly scoots away and hides under adjacent cover. Areas with strong currents seem to suit the shelter-seeking slender cockscomb best, but little is known about many of its other tendencies.

Although too small for marketing, the thin slender cockscomb may sometimes enter fine-meshed prawn traps or become entangled in the web fencing of bait-herring ponds located over rocky bottoms.

South of Alaska, the difficult-to-distinguish slender cockscomb lives at depths inaccessible to intertidal explorers or dockside naturalists. Only along the shallow, rocky shores of Alaska does this species live, with the high cockscomb, intertidally, but even there it is seldom sighted.

Small and rarely obtained, the usually brown, attractively-patterned slender cockscomb is not missed at all, culinarily speaking.

56 Whitebarred Prickleback

SERIES OF WHITE VERTICAL BARS

LARGE
BLUNTLY
ROUNDED
TAIL FIN

SPECIES: *Poroclinus rothrocki*—from the Greek words *poros* and *clinus*, meaning "pore" and "blenny;" as well as honouring J.T. Rothrock, a former University of Pennsylvania botanist.

ALTERNATE NAMES: white-barred blenny.

MAXIMUM RECORDED SIZE: 25 cm (10 inches).

DISTRIBUTION: San Diego, southern California, to Unalaska Island and the Bering Sea coast of Alaska.

It is extremely unlikely that the small, tiny-mouthed whitebarred prickleback would seize a baited hook and actually get caught because bottom-fishermen very seldom use minute hooks when angling in deep water. However, this elongate, active species should prove useful as a live bait for large flounders or codfish if it could be acquired.

While making dives deeper than 17 m (102 feet), the observant diver might conceivably notice the pinkish whitebarred prickleback stretched out and resting upon flat, sandy bottoms. Night divers probably have a better opportunity to observe this slender, easily overlooked creature which, in captivity at least, often hides beneath shells or stones. If you find a whitebarred prickleback, consider yourself a "blue ribbon" fish watcher.

Despite not being a commercially important species which brings the professional fisherman a profit, the light-coloured whitebarred prickleback is not infrequently swept up in fine-meshed shrimp trawl nets towed over soft substrates at depths down to 128 m (422 feet). Like many other small marine fishes, though, this bottom dweller has an indirect, indeterminable value as forage for large, economically significant groundfishes.

A deepwater denizen, the shy whitebarred prickleback does not live around wharves, pilings or jetties, and never becomes trapped in tidepools.

Simply too small and too seldom caught to be of significance to the chef.

57 Pacific Snake Prickleback

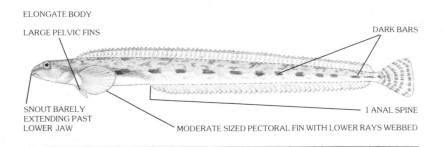

ELONGATE BODY

LARGE PELVIC FINS

DARK BARS

SNOUT BARELY
EXTENDING PAST
LOWER JAW

1 ANAL SPINE

MODERATE SIZED PECTORAL FIN WITH LOWER RAYS WEBBED

SPECIES: *Lumpenus sagitta*—from the German *lumpen*, reference to a somewhat similar European fish; and the Latin *sagitta*, "arrow."

ALTERNATE NAMES: Pacific snakeblenny, eel-blenny.

MAXIMUM RECORDED SIZE: 51 cm (20 inches).

DISTRIBUTION: Humboldt Bay, northern California, to the Aleutian chain, and Bering Sea coast of Alaska, and the Sea of Japan.

During summer bottom-fishermen, particularly those seeking the various flounders over sandy or silty substrates, often unintentionally hook the Pacific snake prickleback. Baits such as marine worms, shrimps, mussels, clams and pieces of fish especially attract this slender, tiny-mouthed creature that can only be caught with small hooks; jetties and piers constructed over flat-bottomed, shallow bays are good angling platforms for it. In spite of its best and most tenacious efforts, the light-weight Pacific snake prickleback provides little sport on any tackle.

A sandy or muddy bottom inhabitant, the seasonally abundant Pacific snake prickleback migrates into shallow coves and inlets during summer and early autumn. By day it is wary and quickly darts off at even a cautious diver's approach, but at night the beam of an underwater flashlight seems to mesmerize or "freeze" the creature, and allow the diver a close look. Unlike many other pricklebacks and gunnels which coil their bodies, this light gray species stretches out, often straight as an arrow.

Shrimp trawlers commonly catch the unwanted Pacific snake prickleback as an unsaleable incidental in shallow summer tows, and less frequently in winter hauls as deep as 207 m (680 feet).

If the water is clear enough, pierside naturalists may notice the ghostly and motionless Pacific snake prickleback on sandy or muddy bottoms. This species swims in quick short bursts interrupted by longer rest intervals. It very rarely frequents tidepools.

Panfry the tasty Pacific snake prickleback for a snack.

58 Rock Prickleback

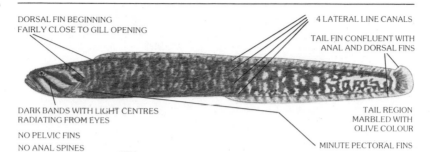

DORSAL FIN BEGINNING
FAIRLY CLOSE TO GILL OPENING

4 LATERAL LINE CANALS

TAIL FIN CONFLUENT WITH
ANAL AND DORSAL FINS

DARK BANDS WITH LIGHT CENTRES
RADIATING FROM EYES

NO PELVIC FINS
NO ANAL SPINES

TAIL REGION
MARBLED WITH
OLIVE COLOUR

MINUTE PECTORAL FINS

SPECIES: _Xiphister mucosus_—from the Greek _ziphister_, meaning "swordbelt;" and Latin _mucosus_, "slimy."

ALTERNATE NAMES: rock blenny, rock-eel, black eel*.

MAXIMUM RECORDED SIZE: 58 cm (23 inches).

DISTRIBUTION: Santa Cruz Island, southern California, to Port San Juan, southeastern Alaska.

Surprisingly, perhaps, some anglers who bottom-fish along shallow, rocky coastlines, particularly in California where bag limits exist for it, take the robust rock prickleback on small hooks baited with pieces of fish or invertebrates. Active, abundant, and available: this elongate prickleback is excellent bait for larger bottomfish too.

The mottled yellow-and-black rock prickleback hides under rocks located in weedy shallows, particularly those along open coast areas. Also look very carefully in large tidepools: it may often just poke its head out from beneath shelter. The herbivorous rock prickleback consumes various marine plants and only ingests animals such as shrimps, bryozoans, sponges, and tiny fishes incidentally with the vegetation.

Not a commercial species, the unmarketable rock prickleback is very rarely caught.

At moderate to very low tides, turn over seaweed-covered rocks when looking for the abundant rock prickleback, a species that may live at least 11 years. Its slippery, rapidly writhing form is so difficult to grasp that those people collecting this fish for food usually do so with a gaff or nail-bearing stick. The robust rock prickleback breeds during winter, and large gravid females, of at least 35 cm (14 inches), deposit their eggs in spherical clusters beneath rocks.

Skeletal remains of the rock prickleback found within ancient middens verify that coastal Indians have eaten it for centuries.

* —incorrect

59 Black Prickleback

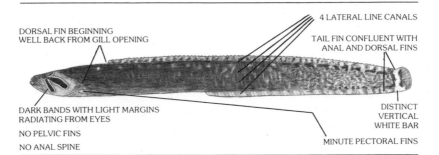

DORSAL FIN BEGINNING
WELL BACK FROM GILL OPENING

4 LATERAL LINE CANALS

TAIL FIN CONFLUENT WITH
ANAL AND DORSAL FINS

DARK BANDS WITH LIGHT MARGINS
RADIATING FROM EYES

NO PELVIC FINS

NO ANAL SPINE

DISTINCT
VERTICAL
WHITE BAR

MINUTE PECTORAL FINS

SPECIES: _Xiphister atropurpureus_—from the Greek _ziphister_, meaning "sword belt;" plus the Latin _ater_ and _purpureus_, meaning "black" and "purple."

ALTERNATE NAMES: black blenny.

MAXIMUM RECORDED SIZE: 30 cm (12 inches).

DISTRIBUTION: Rio Santo Thomas, northern Baja California, Mexico, to Kodiak Island, southeastern Alaska.

The mouth of this fish is too small to take most hooks, but on an intertidal excursion the bottom-fisherman might easily gather a batch of live, easy-to-maintain black pricklebacks before a fishing trip; by occasionally changing the seawater in the bucket of specimens, the angler can be assured of active bait to entice large rockfishes or greenlings.

Sharp-eyed divers exploring shallow rocky shorelines at depths less than 9 m (30 feet) may notice the black prickleback lurking beneath loose stones, often with just its head exposed. Upon overturning the sheltering rock you may notice only a blur as it speeds off to hide beneath other nearby stones.

The small black prickleback inhabits shallow or intertidal rocky locales where commercial fishermen do not operate.

Curious beachcombers, particularly those who brave harsh, exposed Pacific shores, may readily discover the black prickleback by turning over rocks in the low intertidal zone. A very tolerant species, this adaptable fish can survive low salinities created by heavy rainfalls. During late winter and spring, mated pairs of black pricklebacks breed under boulders on rocky or shell substrates. After spawning, the male parent remains for up to three weeks to guard the spherical egg mass; then the tiny larvae hatch and initially swim at the surface of the water. The herbivorous black prickleback feeds upon plants but may also incidentally take in tiny shrimp-like creatures too. It, in turn, is food for creatures such as the garter snake, mink and raccoon.

Although the slender black prickleback may be eaten, it rarely is.

60 Ribbon Prickleback

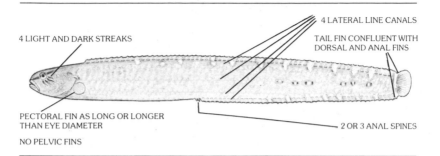

4 LATERAL LINE CANALS

4 LIGHT AND DARK STREAKS

TAIL FIN CONFLUENT WITH
DORSAL AND ANAL FINS

PECTORAL FIN AS LONG OR LONGER
THAN EYE DIAMETER

2 OR 3 ANAL SPINES

NO PELVIC FINS

SPECIES: *Phytichthys chirus*—from the Greek *phytos* and *ichthys*, meaning "plant" and "fish;" plus the Latin *cheir*, "hand."

ALTERNATE NAMES: belted blenny.

MAXIMUM RECORDED SIZE: 20 cm (8 inches).

DISTRIBUTION: southern California, to Adak Island, in the Aleutian chain, and the Bering Sea coast of Alaska.

Any angler fortunate enough to obtain a number of live ribbon pricklebacks will find them to be a good bait for large flounders, rockfishes or greenlings.

Along the open and surf-swept shores, search very closely in large, weedy tidepools or in rocky crevices at depths less than 13 m (43 feet), to discover the elusive ribbon prickleback. Pay particular attention to the bases of large kelp plants where their wide, gnarled holdfasts attach them to the rock: the ribbon prickleback often entwines itself around those holdfasts and its colour may match any of its weedy surroundings.

Because commercial harvesters seek no valuable quarry along the shallow, net-tearing, rocky open coast, they very rarely encounter the unsaleable ribbon prickleback.

Emptying exposed coastal tidepools by bailing is perhaps the best method of gathering the variously coloured ribbon prickleback. Choose a pool very low in the intertidal zone and easily drained by the number of available people with buckets. Try a chain gang! Although a strenuous physical challenge requiring good teamwork, a successful bail may strand many interesting creatures before the incoming tide refills the natural but temporary aquarium. Newly hatched ribbon prickleback larvae live at the surface where they initially feed upon minute animals and plants. Later, larger plants such as red and green algae, in addition to numerous shrimp-like creatures, make up most of the primarily herbivorous ribbon prickleback's diet.

Rarely available, the small ribbon prickleback offers little to the seafood gourmet.

61 Longsnout Prickleback

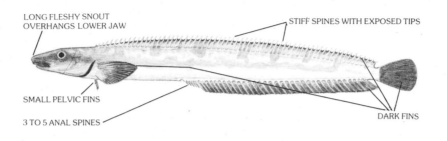

LONG FLESHY SNOUT OVERHANGS LOWER JAW

STIFF SPINES WITH EXPOSED TIPS

SMALL PELVIC FINS

3 TO 5 ANAL SPINES

DARK FINS

SPECIES: *Lumpenella longirostris*—from the German *lumpen*, a name for a somewhat similar European fish; plus the Latin *ella*, *longus* and *rostrum*, meaning "diminutive," "long," and "beak."

ALTERNATE NAMES: none.

MAXIMUM RECORDED SIZE: 31 cm (12.3 inches).

DISTRIBUTION: Howe Sound and Burrard Inlet, southern British Columbia, to Kodiak Island, Gulf of Alaska.

Sport fishermen do not fish the depths where the rarely-seen longsnout prickleback lives, nor fish with hooks tiny enough to catch it. Consequently, this is a species unknown to the sport angler.

Because present records indicate the poorly known longsnout prickleback lives at depths of 92 m (200 feet) or deeper and sport divers seldom descend to half that depth, they understandably have not observed the dark longsnout prickleback. However, as the ever-increasing diving community makes more explorations of virgin territory, this record may yet alter.

The small longsnout prickleback is of no direct economic significance, and remains only an incidental catch for a shrimper towing a fine-meshed trawl net over silty or muddy bottoms at depths between 92 and 141 m (200 and 462 feet). Although this rarely collected prickleback lives as an adult at depths greater than any other members of its family, exploratory scientific samplings of surface plankton tows show that tiny transparent longsnout prickleback larvae swim for the first few weeks of life at the surface where they feed upon tiny shrimp-like animals.

Even if the surface-bound naturalist were to see the clear and sliver-like larva of a longsnout prickleback, that spectator would find the task of distinguishing it from other minute young pricklebacks is practically impossible.

No doubt the slender longsnout is not worth serious consideration as food for people.

THE GUNNELS
(Family: Pholidae)

A very small family of fishes, the gunnels, or Pholidae, contains about thirteen recognizable species, most of which live in temperate North Pacific waters; the several others remaining inhabit the cool or cold waters of the Arctic or North Atlantic. Exclusively shallow marine or even intertidal denizens, the elongate, usually colourful gunnels frequent various bottom habitats where they find secure shelter such as marine plant life, rocky crevices or pilings and sunken logs.

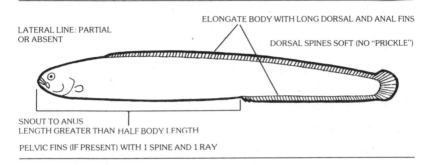

ELONGATE BODY WITH LONG DORSAL AND ANAL FINS

LATERAL LINE: PARTIAL OR ABSENT

DORSAL SPINES SOFT (NO "PRICKLE")

SNOUT TO ANUS LENGTH GREATER THAN HALF BODY LENGTH

PELVIC FINS (IF PRESENT) WITH 1 SPINE AND 1 RAY

Ichthyologists can only guess at the origins of present-day gunnels and at what point their ancestors first appeared because paleontologists have yet to dig up the first definite pholid fossil. Close contemporary relatives of the secretive gunnels include the very similar pricklebacks, wrymouths, and kelpfishes—all elongate fishes with very lengthy dorsal and anal fins at least partially supported by spines. Because no gunnels grow to more than a modest 46 cm (18 inches), they attract no commercial exploitation; only a few photographers, professional aquarists and scientists actively pursue them.

The red gunnel, similar to the species on page 92 but without the light patches along the base of the dorsal fin, is rarely encountered and is therefore not included here.

62 Penpoint Gunnel

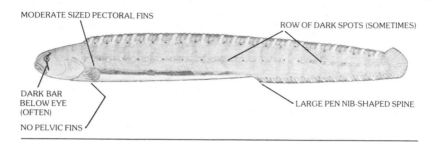

MODERATE SIZED PECTORAL FINS

ROW OF DARK SPOTS (SOMETIMES)

DARK BAR
BELOW EYE
(OFTEN)

NO PELVIC FINS

LARGE PEN NIB-SHAPED SPINE

SPECIES: *Apodichthys flavidus*—from the Greek *apous* and *ichthys*, meaning "without feet" and "fish;" and the Latin *flavidus*, "yellow."

ALTERNATE NAMES: pen-point blenny.

MAXIMUM RECORDED SIZE: 46 cm (18 inches).

DISTRIBUTION: Santa Barbara, southern California, to Kodiak Island, Gulf of Alaska.

The slender, writhing penpoint gunnel attracts hungry rockfishes, greenlings, and flounders, the prime quarries for bottom-fishermen. Because of its tiny mouth, the elongate penpoint gunnel cannot easily take most hook and line apparatus; put on beachcombing boots and look under seaweeds or rocks at low tide for this special bait.

During summer, the penpoint gunnel lives among the seaweed and the rocks. Watch for a green specimen among sea lettuce or eelgrass, the wine-red individual huddling in red algae, and the golden brown in the various kelps. Each colour is directly related to diet. When the transparent larval penpoint gunnel settles to the bottom to feed its individual colour gradually begins to appear, although it may alter with a change in diet associated with other plants. In winter, most of these sheltering plants die and this gunnel then retreats under the rocks.

The seldom-caught penpoint gunnel weaves its way through all but the finest meshed nets—fortunately for both the fishermen, who cannot market the fish and, of course, fortunately for the gunnel itself. In any case, fishermen seldom set their nets in the shallow water where this fish abounds.

Search through seaweeds at low tide in summer for the penpoint gunnel. Locating it is easy, but capturing one is exasperatingly difficult because the slippery creature writhes and wriggles to freedom from between fingers or hands. In late winter, overturn rocks exposed by low tide and perhaps find a penpoint gunnel coiled about a mass of whitish-coloured eggs. Think conservation: carefully replace their shelter.

If unencumbered by preconceptions about size, colour and shape, you just might possibly discover a culinary purpose for large penpoint gunnels.

63 Rockweed Gunnel

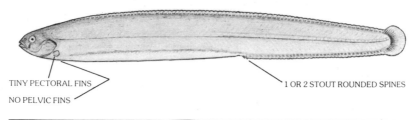

BRIGHT UNIFORM COLOUR

TINY PECTORAL FINS

NO PELVIC FINS

1 OR 2 STOUT ROUNDED SPINES

SPECIES: *Xererpes fucorum* —from the Greek *kseros* and *cerpes*, meaning "dry" and "creeper;" plus the Latin *fucus*, for the algae known as rockweed.

ALTERNATE NAMES: rockweed blenny, fucus blenny.

MAXIMUM RECORDED SIZE: 23 cm (9 inches).

DISTRIBUTION: Punta Escarpada, central Baja California, Mexico, to Banks Island, northern British Columbia.

Though never caught on hook and line, the colourful rockweed gunnel probably makes good bait for large bottomfish such as rockfishes and greenlings.

The intertidal habitat of the brightly-coloured rockweed gunnel allows snorkellers the opportunity to find it in the often murky shallows at high tide or in large, weedy tidepools at low tide. Carefully and deliberately look through marine plants, particularly through the dense, finger-like rockweed growing on craggy rock. Poor visibility along most shores due to incessant wave action or heavy rain often conceals the slender rockweed gunnel from the view of even the most observant diver. This brilliant green or red fish preys upon tiny intertidal animals such as shrimps or snails, and is itself eaten by mink.

Commercial fishermen find no profitable quarry in the intertidal zone including the unexploited rockweed gunnel, one of many small fish remaining unknown to the harvesters of the oceans.

At low tide, inspect closely any exposed marine plants growing on solid rock or overturn weed-covered stones to sight the elusive rockweed gunnel. Once located, though, it quickly slithers away, easily avoiding or slipping between your eager fingers; a fine-meshed dipnet is handy. The creature's habit of lying among weeds that are exposed completely to the air obviously impressed the early collector who named it the "dry creeper."

Many other Pacific Northwest fish species are more worthy of culinary efforts. Try with a grain of salt.

SERIES OF CRESCENT-SHAPED MARKINGS

MINUTE PELVIC FINS EACH WITH 1 SPINE AND 1 RAY

SPECIES: *Pholis laeta*—from the Greek *pholas*, meaning "one who lies in wait;" and the Latin *laeta*, "joyful."

ALTERNATE NAMES: bracketed blenny.

MAXIMUM RECORDED SIZE: 25 cm (10 inches).

DISTRIBUTION: Crescent City, northern California, to the Aleutian Island chain, and the Bering Sea coast of Alaska.

The active crescent gunnel should be useful as a live bait when the angler fishes for rockfishes or ling cod. However, except when inadvertently snagged, this often intertidal gunnel is not caught with conventional tackle.

Shallow subtidal and intertidal locales with various marine algae and eelgrass are prime territory for the distinctively marked crescent gunnel. It often hides under small boulders or in crevices along massive underwater rock formations. This wary creature also often lives around old breakwaters and wharves where it may colonize bottles, jars, cans, tires, or pots, and may stare curiously back at you. Particularly at night a slow-moving and careful diver who does not stir up the bottom may find, in the open, an undisturbed, easily photographed crescent gunnel.

The small crescent gunnel easily evades or escapes from all commercial gear used in fishing the Pacific Northwest.

Especially in summer, the variously coloured crescent gunnel inhabits weed-filled tidepools or lives under encrusted rocks left exposed in the low intertidal zone. A fine-meshed dipnet is a most helpful tool for capturing this squirming, difficult-to-handle gunnel; scoop steadily along the bottom and among the seaweeds. Dockside naturalists may find the common crescent gunnel entwined among the seaweeds or around the marine animals which encrust pilings and the undersides of floats.

The slender and small crescent gunnel is unworthy of culinary consideration.

65 Saddleback Gunnel

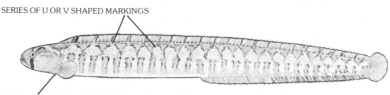

SERIES OF U OR V SHAPED MARKINGS

MINUTE PELVIC FINS EACH WITH 1 SPINE OR 1 RAY

SPECIES: _Pholis ornata_—from the Greek _pholas_, meaning "one who lies in wait;" and the Latin _ornata_, "ornamental."

ALTERNATE NAMES: saddled blenny.

MAXIMUM RECORDED SIZE: 30 cm (12 inches).

DISTRIBUTION: Carmel Beach, central California, to the Bering Sea, and south to Pusan, Korea.

Fortunate indeed is the bottom-fishing enthusiast who captures a number of live saddleback gunnels to use as bait. When the angler puts these active, wriggling creatures on a hook and lowers them into the sea, they trigger voracious responses from hungry rockfishes, greenlings or large-mouthed flounders. Unless unintentionally snagged, though, this gunnel itself is never caught on hook and line.

Shallow muddy substrates densely covered with eelgrass beds are prime locales to dive if searching for the secretive saddleback gunnel. Although this colourful fish exists at depths down to 37 m (120 feet), aquanauts looking for it are most successful when cruising depths much less than half that figure. The olive and orange saddleback gunnel entwines itself among various marine plants or hides beneath scattered rocks or sunken logs.

The slender saddleback gunnel easily weaves its way through nearly any sized mesh of commercial nets that might be placed in the shallow, seldom-harvested habitat of this unmarketable fish.

Particularly during summer, beachcombers may find the slippery saddleback gunnel intertidally under rocks or lurking in muddy-bottomed, eelgrass-filled tidepools. Marine plants hanging from wharves and pilings also shelter this cover-seeking animal but the dockside naturalist must be a patient observer to notice it. In late winter and early spring, a mature female saddleback gunnel deposits small, round egg clusters, a male fertilizes them, and both parents may then remain as guards.

Only a very desperate soul would consider eating the slender saddleback gunnel.

66 Longfin Gunnel

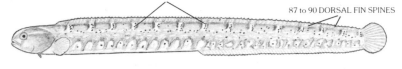

SERIES OF LIGHT MARKS WITH SMALL DARK SPOTS INCLUDED

87 to 90 DORSAL FIN SPINES

SMALL PELVIC FINS

SPECIES: *Pholis clemensi*—from the Greek *pholas*, meaning "one who lies in wait;" and honouring Wilbert A. Clemens, a legend in Canadian ichthyology.

ALTERNATE NAMES: none.

MAXIMUM RECORDED SIZE: 13 cm (5 inches).

DISTRIBUTION: Arena Cove, Mendocino County, northern California, to Fillmore Island, southeastern Alaska.

The little but active longfin gunnel, with its often bright coloration, is potentially a fine live bait, but the average bottom-fishing angler remains unlikely to obtain enough specimens for a serious fishing session.

The tiny longfin gunnel may stare back intently from among branched leafy red algae or the adjacent rocks. In shallow to moderately deep reef locales, concentrate on depths below 8 m (27 feet) and watch closely for undisturbed specimens which may be resting on exposed rock formations before they quickly retreat to sheltering crevices. Most Pacific Northwest fish species were described in the 19th century or before; amazingly, the common longfin gunnel was only discovered in 1964. Because at present much of its biology, ecology and distribution remains unknown, divers will contribute greatly to this knowledge because they alone may study and observe this creature as it "lies in wait" in its natural surroundings. Such an observation was made in January of 1985 when underwater photographer Bernie Hanby photographed an adult longfin gunnel wrapped around a circular clutch of white eggs in a chimney sponge at about 30 m in the Pender Harbour area, British Columbia. The observation of a longfin gunnel resting in the slightly agape jaws of a large lingcod suggests that this attractive gunnel sometimes picks parasites from larger, tolerant fish in a special relationship.

The unsaleable longfin gunnel can easily slip through most commercial nets, but lives in inaccessible rocky habitats, at depths to 64 m (210 feet), where few commercial fishermen harvest anyway.

Only recently discovered, the delicate-looking longfin gunnel may live intertidally, and perhaps knowledgable beachcombers may verify that it does.

The longfin gunnel is too small and too thin to eat.

THE WOLFFISHES
(Family: Anarhichadidae)

The small family of robust fishes, scientifically called the Anarhichadidae are actually of two distinguishable subgroups. Most of the family is collectively called "wolffishes" and contains approximately eight known living species of the North Atlantic, all of which are only moderately elongate: they are stubby creatures with distinct, large tail fins. Living in the Pacific Northwest, the wolf-eel, with its very long, slender body that tapers into a pointed, barely discernible tail fin, is, by itself, one subgroup of the wolffish family. All anarhichadids live upon rocky reef or stony shelving bottoms at shallow to moderate depths, and stake out territories that usually include a crevice, den or lair.

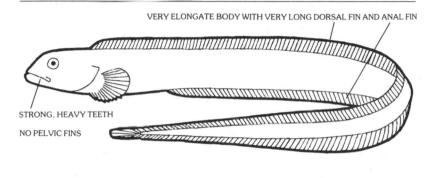

VERY ELONGATE BODY WITH VERY LONG DORSAL FIN AND ANAL FIN

STRONG, HEAVY TEETH

NO PELVIC FINS

Anarhichadids, which all grow to at least 91 cm (36 inches), have strong, powerful jaws with thick sharp spike-like front teeth, and flat, heavy molars. These impressively armed mouths enable the large wolffishes to prey upon many hard-shelled animals such as clams, snails, and crabs. These brutes are dangerous when under stress: dangling from a hook, enmeshed in a net, or cornered in a trap.

Unfortunately, no fossil anarhichadids are yet known to science and so detail about wolffish origins remain a mystery. Of the world's large and varied fish population, though, experts believe that pricklebacks, gunnels and wrymouths relate most closely with modern wolffishes; all have long dorsal fins supported by spiny, unsegmented rays. Throughout the North Atlantic, the large wolffishes are significant commercial species for trawlers who take them incidentally. Not only is the flesh of anarhichadids very tasty, but also the tanned skin makes useful leather.

67 Wolf-Eel

DARK SPOTS SURROUNDED BY LIGHT HALOS

VERY ELONGATE BODY

LARGE CANINE AND MOLAR TEETH

NOTCHES

NO PELVIC FINS

SPECIES: *Anarrhichthys ocellatus*—from the Greek *anarrhichas*, an ancient name for a similar fish; and the Latin *ocellatus*, "eye-like."

ALTERNATE NAMES: none.

MAXIMUM RECORDED SIZE: 2.4 m (8 feet) and 18.4 kg (40.6 pounds).

DISTRIBUTION: Imperial Beach, southern California, to the Krenitzen Islands, in the Aleutian chain, Alaska. Also reported from the Seas of Japan and Okhotsk.

Although not usually sought, large wolf-eels sometimes surprise anglers fishing rocky bottoms and using baits such as anchovy, abalone, or crab. Exercise *extreme caution* when removing the hook from the strong mouth of a writhing wolf-eel because its heavy, teeth-laden jaws could inflict a savage and dangerous wound!

A very common inhabitant of the rocky reef and shoreline, the gray wolf-eel most often lurks within a cave or crevice. A lucky aquanaut may notice a cavern-dwelling mated pair, the whitish, puffy-headed male being easily distinguished from the darker female. Underwater photographers delight in locating this tameable species and often, by gentle hand feeding, usually with red sea urchins, coax it from its lair. Such "buddies" readily swoop out, slowly and gracefully, to greet all approaching divers.

Although sometimes filleted and sold fresh, the incidentally trawled wolf-eel is really economically insignificant. Other fishing methods too, including longlines, handlines and traps for prawns, crabs or salmon, occasionally capture wolf-eels from depths to 226 m (740 feet).

The impressive wolf-eel dwells too deep to be sighted by surface-bound naturalists.

Historically, some coastal Indians prized the wolf-eel as the "doctorfish" or *mukah*; only the medicine man ate the tasty, white flaky flesh in order to enhance his healing powers.

THE WRYMOUTHS
(Family: Cryptacanthodidae)

A very small family of fishes currently containing only four recognizable living species, the Cryptacanthodidae translates to mean "hidden spine family," and are called wrymouths. Intertidal domains, as well as shallow to moderately deep habitats in both the North Atlantic and North Pacific Oceans, provide territory for these few elongate, homely-looking creatures. Exclusive residents of sandy, muddy or silty bottoms, these little-known fishes spend nearly all their time buried in the sea floor, often within extensive subterranean tunnels. Lurking here while awaiting prey or mates, and avoiding predators, the inactive cryptacanthodids only occasionally venture out to prowl slowly over the seascape.

In spite of years of effort, collections of ancient wrymouth remains are very scanty but from these sparse fossils, paleontologists know cryptacanthodids probably originated 30 million years ago during Eocene times. Their closest living relatives are among the elongate blenny fishes of northern seas, which include the gunnels, pricklebacks and wolffishes; common characteristics of all include very long, low dorsal fins supported by hard, spiny rays. Although several of the wrymouth species attain large sizes, up to 117 cm (46 inches), no fisheries exist for them, and the few incidentally trawled are either sold to reduction plants or discarded at sea.

FLATTENED HEAD WITH
UPWARDLY DIRECTED MOUTH

ELONGATE BODY, WITH LONG DORSAL AND ANAL FINS

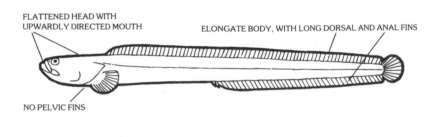

NO PELVIC FINS

Sometimes large wrymouths become hooked on a longline, a specific and efficient type of commercial fishing gear sometimes called a setline. Fished over the sea floor, a longline consists of a long primary or main line to which, at intervals of about 3 m, up to several hundred 2 m or shorter lengths of line are attached, each shorter length having a baited hook at its other end. Anchors attached at each end of the main line sink the gear to the bottom and two other "float" lines, one at each end of the primary line, carry buoys to mark this gear at the surface. Commercial fishermen may leave this rig out for several hours before winching it aboard to remove and sort the catch, which may include a large wrymouth.

68 Giant Wrymouth

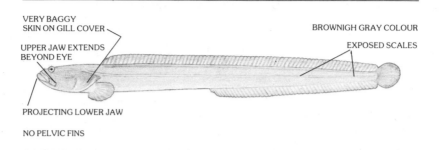

VERY BAGGY
SKIN ON GILL COVER

BROWNIGH GRAY COLOUR

UPPER JAW EXTENDS
BEYOND EYE

EXPOSED SCALES

PROJECTING LOWER JAW

NO PELVIC FINS

SPECIES: *Delolepis gigantea*—from the Greek *delos* and *lepis*, meaning "visible" and "scale;" plus the Latin *gigantea*, "huge."

ALTERNATE NAMES: potatohead, congo eel*, buckskin eel*.

MAXIMUM RECORDED SIZE: 117 cm (46 inches).

DISTRIBUTION: Humboldt Bay, northern California, to Unalaska Island, in the Aleutian chain, Alaska.

Every once in a while a startled salmon fisherman "mooching" with herring, or a shocked bait-fisherman offering cut fish to bottom-dwelling flounders or codfishes accidentally hooks a lethargic giant wrymouth. A flabby and slow fish, it offers little resistance once hooked. Such captures occur infrequently, probably because this elongate, secretive fish spends most of its life buried in the silty or sandy bottom.

Not only does the homely giant wrymouth usually prefer soft, level bottoms below 20 m (66 feet), a zone seldom explored by sport divers, but this sandy-coloured creature also spends much of its time buried, with only its large gnarled head occasionally exposed to the diver. Underwater photographer Neil McDaniel, an avid naturalist, found and "shot" several such subjects off the mouth of the Fraser River, British Columbia.

While seeking valuable bottom-dwelling seafoods, commercial fishermen once every so often catch the unmarketable giant wrymouth: trawlers inadvertently scoop it with nets dragged along the bottom, longliners accidentally take it with their multi-hook set lines, and surprised crab trappers or prawn potters periodically find it writhing awkwardly in their cages.

Once in a while the unusual giant wrymouth ventures into shallow silty bays where it may rest motionless on the bottom, often appearing to be dead. Sometimes, also, a bemused beachcomber will find a specimen stranded in a shallow tidepool choked with eelgrass near the water's edge.

If the chef can overcome an initial repulsion, there is a large quantity of edible flesh on an adult giant wrymouth. Remember, beauty is only skin deep!

*—incorrect

69 Dwarf Wrymouth

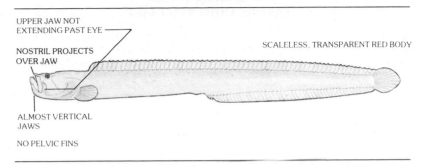

UPPER JAW NOT
EXTENDING PAST EYE

NOSTRIL PROJECTS
OVER JAW

SCALELESS, TRANSPARENT RED BODY

ALMOST VERTICAL
JAWS

NO PELVIC FINS

SPECIES: *Lyconectes aleutensis*—from the Greek *lycos* and *nectes*, meaning "wolf" and "swimmer;" plus emphasizing the Aleutian Islands.

ALTERNATE NAMES: red devil.

MAXIMUM RECORDED SIZE: 30 cm (12 inches).

DISTRIBUTION: Eureka, northern California, to the Aleutian Islands, and the Bering Sea, Alaska.

Its subterranean habits and small size isolate the bright red dwarf wrymouth from most anglers. If available, though, it would make very tempting bait for codfishes or large flounders.

The few divers who patrol over silty bottoms at depths greater than 30 m (100 feet) may unknowingly cruise over many hidden dwarf wrymouths because this bright red fish burrows into soft substrates completely obscuring all of its scaleless and wrinkled body.

The small dwarf wrymouth is of no direct commercial value, but occasionally the spreader "doors" of a fine-meshed shrimp trawl dig deeply into soft, muddy bottoms, at depths down to 350 m (1,150 feet), dislodge a startled specimen and sweep it into the net. In the northern parts of its geographic range, the beady-eyed dwarf wrymouth tends to live in shallower water than it does in the south—a phenomenon common to many bottomfish species of the Pacific shores of North America.

Unfortunately unidentifiable by all but a few experts, transparent, elongate dwarf wrymouth larvae, less than 3 cm long, attracted by dock lights, frequently swarm at the surface during late spring and early summer. As temporary members of the planktonic community, they feed upon fellow floating creatures, notably copepods. Weeks later these dwarf wrymouth larvae develop distinctive adult characteristics and descend to a permanent bottom dwelling existence.

Perhaps some enterprising chef will obtain enough dwarf wrymouth for experimentation.

97

THE QUILLFISHES
(Family: Ptilichthyidae)

One of tiniest families of fishes on this planet, the quillfish family contains but a single known species, which lives in the Pacific Northwest. A highly modified creature of which very little is known, the quillfish is most closely related to the various gunnels, pricklebacks and wolffishes. Not unexpectedly, because of its remarkable thread-thin form, no fossilized ptilichthyids have yet been unearthed.

VERY LONG SLENDER BODY

TINY HOOKED SPINES

TINY MOUTH
WITH PROTRUDING LOWER JAW

Undoubtedly the best way for the wharf or float-bound naturalists to view the strange, unbelievable quillfish is by watching closely at a "night light,"—a watertight fixture hung just below the surface of the sea. After purchasing an outdoor light fixture with a plastic or glass cover, seal it with melted wax to prevent leakage. There are two possibilities for power: a 110-watt system which requires an extension cord for power and a convenient socket, or a 12-volt circuit wired to a fully charged automobile battery. With either, *beware of shock* when night-lighting around salt water.

Although sighting a quillfish at the night light is an extremely rare and thrilling event, a multitude of creatures of all shapes and sizes are attracted to the glow. Once establishing a handy locale, night light through all seasons to see the great diversity of marine life flourishing at the surface. Usually the wharves or piers swept by slow currents are the best places to see animals that actively swarm to the night light as well as those, such as the fragile and ornate jellyfishes, which passively drift by.

70 Quillfish

VERY LONG, SLENDER BODY

TINY HOOKED SPINES

TINY MOUTH
WITH PROTRUDING LOWER JAW

THREAD-LIKE TAIL
(OFTEN MISSING)

SPECIES: *Ptilichthys goodei*—from the Greek *ptilon* and *ichthys*, meaning "quill" and "fish;" as well as honouring G.B. Goode, a distinguished American ichthyologist.

ALTERNATE NAMES: quill-fish.

MAXIMUM RECORDED SIZE: 34 cm (13.5 inches).

DISTRIBUTION: Oregon, to the Bering Sea coast of Alaska, through the Aleutian chain, to the Okhotsk Sea and the Kurile Islands, USSR.

The amazing quillfish is prey to various fishes, including the popular coho salmon, but its thread-like body proves virtually impossible to affix upon a hook.

Only right at the surface might the most observant diver notice the rarely seen quillfish swimming with snake-like undulations of its pencil-lead-thick body. Perhaps the most plausible occasion would be at night while the diver snorkels at the surface in preparation for descent.

Obviously not a commercially exploited species, the yellowish-brown quillfish might occasionally swim at the surface where nocturnally active fishermen may notice the creature weaving its way through their nets.

In certain localities such as Friday Harbour, Washington, float-bound observers sometimes notice the thin, mirage-like quillfish meandering slowly and gracefully at the surface. Sightings seem to occur at night and particularly in late spring or early summer when this amazing creature appears around light beams shining into the sea. Dr. Charles Moffett, a psychiatrist by profession, assisted with the collection of several quillfish, one of which appears in the photograph within this volume. Dr. Moffett has sighted 15 live specimens in 13 years of night-lighting—undoubtedly a world's record. Speculation persists about where the fragile-looking quillfish spends its daylight hours and if it is primarily an open seas animal only occasionally swept inshore.

The very idea of preparing a quillfish for the dinner table is incomprehensible!

THE GRAVELDIVERS
(Family: Scytalinidae)

Containing only a single known living species, the graveldivers, scientifically termed Scytalinidae, is therefore one of the tiniest families of fishes in the world. That paleontologists, after many years of searching and digging the earth's crust, have not unearthed any recognizable scytalinids, indicates that this little-known group probably never was much larger. In spite of consisting of only one species, the unusual graveldivers are a very distinct group with numerous special anatomical and behavioural features dictating that they rank as a family. Such single species families are termed "monotypic."

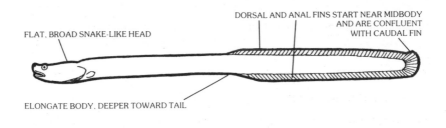

DORSAL AND ANAL FINS START NEAR MIDBODY
AND ARE CONFLUENT
WITH CAUDAL FIN

FLAT, BROAD SNAKE-LIKE HEAD

ELONGATE BODY, DEEPER TOWARD TAIL

When first encountered by most people, scientific names such as the graveldiver's *Scytalina cerdale* appear baffling, strange and perhaps unnecessary. However, they are in fact, the best and most universal names available for defining fishes and may be pronounced simply by using standard English language phonetics: *Scytalina cerdale* sounds like *sky-tal-eena sur-da-lee*. Scientific names prove most useful when someone is communicating to people speaking another language because these scientific names are universal and eliminate completely the vagaries of regional alternatives. These definitive binomials, usually derived from Greek or Latin, correspond to names of ordinary people except that the first element in the scientific set equates with the personal family name and the second element with the given name. *Scytalina cerdale* equates with Smith, John. Readers should become familiar with the scientific names because they can always find information concerning a certain species under that creature's scientific name that otherwise might be unavailable.

71 Graveldiver

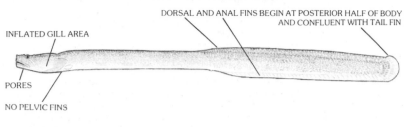

INFLATED GILL AREA

DORSAL AND ANAL FINS BEGIN AT POSTERIOR HALF OF BODY AND CONFLUENT WITH TAIL FIN

PORES

NO PELVIC FINS

ELONGATE, ROUNDED SCALELESS BODY

SPECIES: *Scytalina cerdale*—from the Greek *scytalina* and *cerdale*, meaning "small viper" and "wary one."

ALTERNATE NAMES: burrowing blenny.

MAXIMUM RECORDED SIZE: 15 cm (6 inches).

DISTRIBUTION: Diablo Cove, central California, to Agattu Island, in the Aleutian chain, and the Bering Sea coast of Alaska.

Because this small, thin fish always lives in the gravel, it never takes a hook, but its active elongate body makes excellent live bait for large bottomfishes.

SCUBA divers never see the tiny, inconspicuous graveldiver because it never emerges from its intertidal, under-gravel habitat. Trying to dig them out under water would be futile because the fish wiggles ever deeper while the diver destroys his own visibility.

Nobody places any direct economic value upon the wary graveldiver and only a few harvesters exploit its intertidal habitat. Even commercial clam diggers very rarely dislodge and see this fish because it prefers very loose, relatively clean gravel to the sand where valuable bivalves live.

Beachcombers who stroll along pebbly, open coast beaches are the most likely to locate the secretive graveldiver. At low tide, look along loose gravel beaches or drain tidepools with pebble or broken shell bottoms. Carefully dig into the small stones, either with a shovel or your hands, and watch very closely for these well-camouflaged creatures squirming deeper into the remaining gravel. Once noticing specimens, work as a team: one person shoveling scoops of gravel into plastic basins or buckets and the others sorting through the partially-filled containers for trapped graveldivers. Most of these fish mature at three years of age and some may survive at least another six.

Perhaps a novel Pacific Northwest hors d'oeuvre might be smoked graveldiver!

THE SCORPIONFISHES
(Family: Scorpaenidae)

Among the three hundred and thirty or more species composing the large worldwide family called Scorpaenidae are the tropical stonefishes, scorpionfishes, and lionfishes, as well as the North Pacific rockfishes and thornyheads. The strictly marine scorpaenids prefer shallow to moderately deep habitats and flourish particularly in tropical latitudes such as the Indo-Pacific. Many of the warm-water species nearly always remain motionless, huddling in crevices and crannies of coral reefs, whereas many rockfish species of the cold northern waters swim actively about in rocky regions of the continental shelves, or school well above the uneven bottom. Incidentally, the largest Pacific Northwest rockfishes rank as the giants of the scorpionfish family. Sport anglers seeking the various species of Pacific Northwest rockfishes should be aware of local bag limits imposed in some regions.

An excellent name "scorpionfishes," because the spines of these creatures carry sacks of venom. The best known are the ornate and beautiful but dangerous lionfishes of the tropics. Fortunately for the people of the Pacific Northwest, the venom of their local, temperate water rockfishes is not so potent, though the sharp, strong fin spines may inject a mildly toxic fluid. Exercise caution when handling these fishes, for a jab from their fin spines may cause a throbbing and burning pain, swelling, and even a fever.

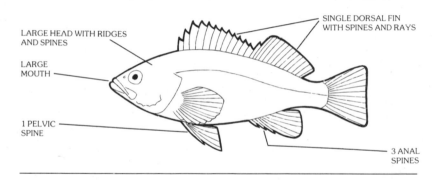

LARGE HEAD WITH RIDGES
AND SPINES

LARGE
MOUTH

1 PELVIC
SPINE

SINGLE DORSAL FIN
WITH SPINES AND RAYS

3 ANAL
SPINES

Many of the Pacific Northwest rockfishes live in very deep water or are inaccessible to most people: these include the rougheye rockfish, the Pacific Ocean perch, the aurora rockfish, the redbanded rockfish, the shortraker rockfish, the darkblotched rockfish, the chilipepper, the rosethorn rockfish, the shortbelly rockfish, the greenspotted rockfish, the grass rockfish, the yellowmouth rockfish, the stripetail rockfish, the harlequin rockfish, the pygmy rockfish, the sharpchin rockfish, and the longspine thornyhead.

72 Copper Rockfish

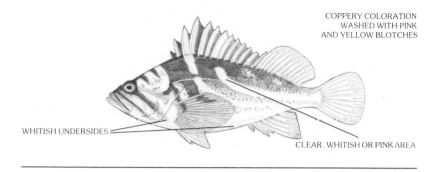

COPPERY COLORATION
WASHED WITH PINK
AND YELLOW BLOTCHES

WHITISH UNDERSIDES

CLEAR. WHITISH OR PINK AREA

SPECIES: *Sebastes caurinus*—from the Greek *sebastos*, meaning "magnificent;" and the Latin *caurinus*, "northwest."

ALTERNATE NAMES: yellow-backed rockfish, white rock cod*, rock cod*.

MAXIMUM RECORDED SIZE: 57 cm (22.5 inches).

DISTRIBUTION: Central Baja California to the Kenai Peninsula, Gulf of Alaska.

A plentiful and common inshore rockfish popular with many light-tackle anglers fishing from wharves and jetties, along rocky shores, or over shallow reefs. It prefers baits—particularly small silvery fishes, marine worms or shrimps—but will also strike at slow-moving lures. Once hooked, the initially feisty copper rockfish swims for cover, sometimes breaking the line on the rocks.

Particularly abundant in shallow, protected bays and inlets where it huddles among rocks or kelp beds, the usually non-schooling adult copper rockfish often rests propped by its fins on the bottom. While primarily foraging near the bottom, this colourful fish will periodically rise into mid-water for such prey as herring, sand lance, anchovies, and shiner seaperch. Divers may also notice the adult copper rockfish lurking around pilings and jetties or under floats and, in summer, tiny juveniles—released live by adult females in April and May—swarming among eelgrass and kelp.

Forming only a minor part of the total "rockfish" catch, the reddish-brown copper rockfish sometimes becomes entrapped in trawls but more often seizes set line hooks. It sells primarily as fresh fillets or live in Chinese restaurants and fish markets.

Look for juvenile copper rockfish loosely aggregated in shallow weedy bays, around wharves or even among the floating drift associated with summer tidelines. Tidepools often temporarily trap young specimens.

Typical of most rockfish, copper rockfish flesh is tasty, firm and flaky: excellent for fish and chips or for panfrying as fillets.

*—incorrect

73 Quillback Rockfish

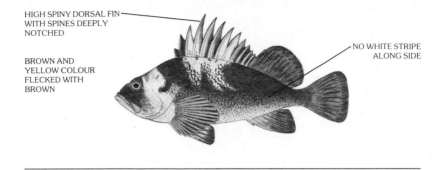

HIGH SPINY DORSAL FIN
WITH SPINES DEEPLY
NOTCHED

BROWN AND
YELLOW COLOUR
FLECKED WITH
BROWN

NO WHITE STRIPE
ALONG SIDE

SPECIES: _Sebastes maliger_—from the Greek _sebastos, malus_ and _gero_, meaning "magnificent," "mast" and "to bear."

ALTERNATE NAMES: speckled rockfish, orange-spotted rockfish, yellow-backed rockfish, brown rockfish*, gopher rock cod*, rock cod*.

MAXIMUM RECORDED SIZE: 61 cm (24 inches).

DISTRIBUTION: Point Sur, central California, to Prince William Sound, Gulf of Alaska.

 Very commonly caught by keen bottom-fishermen with baits such as herring, shrimp and marine worms, or with artificial lures and jigs, the common quillback rockfish frequently annoys salmon-seeking anglers who "mooch" with whole herring. Although wharf-bound, light-tackle anglers may take it, people in boats most often catch this lethargic species over rocky reefs at depths below 15m (60 feet). Handle the hooked quillback rockfish with care, and especially avoid its deeply notched and sharp dorsal fin spines.

 The abundant, usually non-schooling quillback rockfish rests on the rocky bottom or hovers just above it, but never far from cover. Found at all diveable depths, this photogenic species often poses artistically within large cloud sponges, chimney sponges or rocky caverns. Nearly all Pacific Northwest spearfishermen bag a large quillback rockfish or two.

 Forming a significant amount of the rockfish trawl catch, the speckled quillback rockfish is primarily dragged from hard, even bottoms of inlets in water less than 275 m (900 feet), then marketed as fillets. Shrimp trawlers and longliners, too, incidentally but commonly harvest lighter-coloured specimens. The female quillback rockfish usually grows larger than the male.

 Barely recognizable, finely spotted juvenile quillback rockfish less than 2 cm long frequently hide among the floating debris of summer tidelines, but rarely in tidepools.

 Particularly good when deep fried, quillback rockfish is delicious in nearly any recipe!

*—incorrect

74 Brown Rockfish

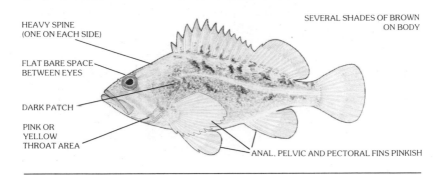

HEAVY SPINE
(ONE ON EACH SIDE)

SEVERAL SHADES OF BROWN
ON BODY

FLAT BARE SPACE
BETWEEN EYES

DARK PATCH

PINK OR
YELLOW
THROAT AREA

ANAL, PELVIC AND PECTORAL FINS PINKISH

SPECIES: *Sebastes auriculatus* —from the Greek *sebastos*, meaning "magnificent;" and the Latin *auriculatus*, "eared."

ALTERNATE NAMES: bolina, brown rock cod*, rock cod*.

MAXIMUM RECORDED SIZE: 55 cm (21.5 inches).

DISTRIBUTION: Bahia San Hipolito, central Baja California, Mexico, to southeastern Alaska.

Most frequently taken from wharves, jetties or along shallow rocky shorelines, the slow-swimming brown rockfish is abundant in certain areas, but scarce in many others. Try bait such as pieces of fish, small crabs or marine worms, and use light spinning tackle to take advantage of the modest sporting value of this fish. If successful, be wary when handling your catch, because its sharp fin spines are mildly venomous and can cause annoying wounds. Small, live specimens make excellent bait for large lingcod.

One of the more sedentary rockfishes, this species, with its several beautiful brown tones, lurks among sheltering weed-covered rocks or around pilings, both easily visited by the diver. This approachable fish is particularly noticeable in Lower Puget Sound. Spearfishermen find the brown rockfish an easy target, as does the underwater photographer, particularly if the water is clear.

Commercial trawlers seldom risk their gear in the shallow rocky habitat of the moderate-sized brown rockfish and professional Pacific Northwest line fishermen only rarely catch it as an incidental. Consequently, little of its flesh arrives at the marketplace.

If the water is clear enough, the studious dockside observer may see the dark form of the brown rockfish but will be unable to identify it specifically from such a vantage point. Almost never found in tidepools.

Fillet the firm, flavourful brown rockfish, then enjoy it in any recipe requiring flaky, white fish. Deep fried whole in peanut oil and glazed with brown sugar it is a favorite of Chinese North Americans.

*—incorrect

105

75 China Rockfish

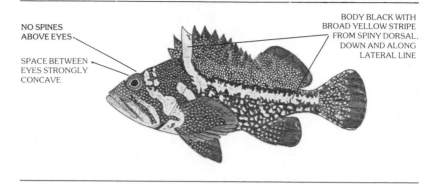

NO SPINES
ABOVE EYES

SPACE BETWEEN
EYES STRONGLY
CONCAVE

BODY BLACK WITH
BROAD YELLOW STRIPE
FROM SPINY DORSAL,
DOWN AND ALONG
LATERAL LINE

SPECIES: *Sebastes nebulosus*—from the Greek *sebastos*, meaning "magnificent;" and the Latin *nebulosus*, "clouded."

ALTERNATE NAMES: yellowstriped rockfish, yellowstripe rockfish, yellow spotted rockfish, Chinese rockfish, yellow rockfish, black and yellow rockfish *, gopher *, china rockcod *, rockcod *, bass *.

MAXIMUM RECORDED SIZE: 43 cm (17 inches).

DISTRIBUTION: San Miguel Island and Diablo Cove, southern California, to southeastern Alaska.

A very common inhabitant of the outer coast, the distinctively coloured china rockfish is most accessible to bottom-fishermen angling from boats. Use baits such as squid, shrimp and small live fishes, or even silvery jigs, at depths less than 25 m (80 feet). Be alert for a "strike." Then hoist the fish toward the surface before this shelter-seeker rushes for cover and makes a line-snapping escape. Handle with care and avoid its sharp and mildly venomous spines.

While patrolling shallow, rocky shores or reefs directly exposed to Pacific surge, an aquanaut often finds the territorial and solitary china rockfish. Curious and unafraid, this beautiful fish often boldly confronts and stares back at the diver, making itself extremely popular with underwater photographers. Take advantage of its head-on pose for nicely framed photos, but bracket exposures to avoid disappointment.

China rockfish live no deeper than 126 m (420 feet) and very few reach market because draggers seldom trawl their easily torn nets over the precipitous rocky habitat of this species.

It is very unlikely, though not impossible, that a surface-bound naturalist will see the black china rockfish, with its indistinct yellow stripes and fine blue speckling.

The small china rockfish's short fillets are worth panfrying or deepfrying any time.

* —incorrect

76 Yellowtail Rockfish

GREENISH OR YELLOWISH
COLOUR—PARTICULARLY ON FINS

SPACE BETWEEN EYES
CONVEX WITH SPINES
THERE EITHER WEAK
OR ABSENT

PROMINENT
KNOB ON END
OF CHIN

POSTERIOR EDGE OF
ANAL FIN VERTICAL

SPECIES: *Sebastes flavidus*—from the Greek *sebastos*, meaning "magnificent;" and the Latin *flavidus*, "yellow."

ALTERNATE NAMES: yellow-tail rockfish, yellow-tail rock-fish, yellowtail rock cod*, yellowtail rockcod*, green snapper*.

MAXIMUM RECORDED SIZE: 66 cm (26 inches).

DISTRIBUTION: San Diego, southern California, to Kodiak Island, Gulf of Alaska.

When congregating near the surface in large, loosely formed schools, scrappy yellowtail rockfish provide excellent sport for the light tackle angler who casts lures or still-fishes with bait. While seeking their prestigious quarry, salmon anglers occasionally catch the swift but disdained yellowtail rockfish as an incidental. In some locales, pier-bound anglers readily take small specimens by day or even at night under bright dock light.

Often associating with other rockfish species, the active yellowtail rockfish congregates in open water along steeply sloping shores or above rocky reefs, but at other times it holes up amid the cracks and crevices of the sea floor. Diving studies connected with capture-release-recapture research indicate that a displaced adult yellowtail rockfish may find its way back, many miles, to its original "home" even after three months. Excellent underwater photos of yellowtail rockfish against a dark background are possible if the water is clear and the exposures are correct.

Trawling and longlining, at depths down to 274 m (900 feet) account for a significant commercial yellowtail rockfish catch which is filleted, then sold, either fresh or frozen, with other rockfishes. In winter one large pregnant female may release as many as 633,000 tiny, living young.

Look for yellowtail rockfish either as juveniles swarming around floats and pilings, or as adults splashing after prey at the surface.

Best when filleted and quickly chilled, as are other rockfishes.

*—incorrect

77 Black Rockfish

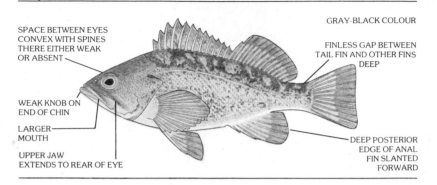

SPACE BETWEEN EYES CONVEX WITH SPINES THERE EITHER WEAK OR ABSENT

GRAY-BLACK COLOUR

FINLESS GAP BETWEEN TAIL FIN AND OTHER FINS DEEP

WEAK KNOB ON END OF CHIN

LARGER MOUTH

UPPER JAW EXTENDS TO REAR OF EYE

DEEP POSTERIOR EDGE OF ANAL FIN SLANTED FORWARD

SPECIES: *Sebastes melanops*—from the Greek *sebastos*, *melas* and *ops*; meaning "magnificent," "black" and "face."

ALTERNATE NAMES: black rock-fish, bass rockfish, black bass*, black seabass*, black rock cod*, black snapper*, sea bass*, bluefish*.

MAXIMUM RECORDED SIZE: 61 cm (24 inches) and 4.8 kg (10.5 pounds).

DISTRIBUTION: Paradise Cove, northern Baja California, Mexico, to Amchitka Island, in the Aleutian chain, Alaska.

One of the gamest of the rockfishes, the very common, schooling black rockfish eagerly pursues artificial lures such as jigs, spoons, and even large flies, but may also take bait. Fish near the surface, over kelp-covered reefs, or along rocky shores to catch the largest specimens. Even at night under artificial lighting anglers may find small ravenous individuals difficult to avoid. Whether trolling or mooching, salmon seekers may take them incidentally as well.

The diver often sees many large black rockfish, often together with other rockfish species, hovering in the green water above, but may also notice specimens huddled among nearby boulders and crevices. Although spearfishermen find large black rockfish an inviting open-water target, underwater photographers may have trouble determining optimal exposures with which to film this dark fish.

The popular black rockfish comprises a variable part of the yearly commercial rockfish harvest, with quantities being handlined, longlined, or trawled at depths to 360 m (1200 feet). Disgruntled salmon trollers often encounter it too, but disdainfully discard it instead of filleting and marketing it.

Easily viewed in summer by the naturalist, young streamlined black rockfish, about 5 cm long, and each with a noticeable black spot on the dorsal fin, swarm abundantly around wharves and pilings, or become trapped in tidepools. Adults commonly "break water" when after prey.

Excellent.

*—incorrect

78 Blue Rockfish

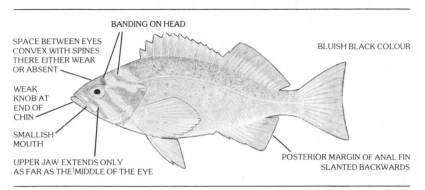

BANDING ON HEAD

SPACE BETWEEN EYES
CONVEX WITH SPINES
THERE EITHER WEAK
OR ABSENT

BLUISH BLACK COLOUR

WEAK
KNOB AT
END OF
CHIN

SMALLISH
MOUTH

UPPER JAW EXTENDS ONLY
AS FAR AS THE MIDDLE OF THE EYE

POSTERIOR MARGIN OF ANAL FIN
SLANTED BACKWARDS

SPECIES: *Sebastes mystinus*—from the Greek *sebastos* and *mystas*, meaning "magnificent" and "priest."

ALTERNATE NAMES: priestfish, priest-fish, black rockfish*, black rockcod*, black bass*, blueperch*, blue perch*, bluefish*.

MAXIMUM RECORDED SIZE: 53 cm (21 inches).

DISTRIBUTION: Punta Blanca, northern Baja California, Mexico, to the Bering Sea coast of Alaska.

Particularly for Californian party-boat anglers, the active, schooling blue rockfish is a very important open coast gamefish that usually swims well off the bottom near kelp growth or over reefs. While spincasting near the surface is an enjoyable way to catch the blue rockfish, still-fishing with small silvery fishes or squid as bait may also be rewarding.

At first visible only as dark shapes shrouded in the greenish water overhead, wary blue rockfish may eventually cruise nearer and nearer to the diver. Because it only occasionally cowers among boulders or cliff-face crevices, eager spearfishermen usually find it a challenging, open-water target requiring patience and careful stalking. Look among lush kelp growths in less than 10 m (30 feet) to find the juvenile blue rockfish, often recognizable by its light blue body spotted with brick red.

Californian longliners and handliners extensively harvest blue rockfish, fillet it, then generally sell it fresh, but trawlers in the north take it only in modest amounts and market it with other rockfishes.

Along surf-swept outer coasts, receding tides very occasionally trap juvenile blue rockfish in weed-choked, rocky pools accessible to adventurous beachcombers. Boaters may occasionally notice adults of this species leaping clear of the water while pursuing small fishes.

Broiled, baked, pan fried, deep fried, boiled, barbecued, or even raw with soy sauce as *sashimi*, the firm blue rockfish tastes delicious.

*—incorrect

109

79 Dusky Rockfish

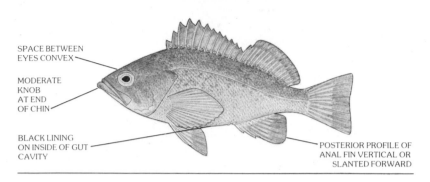

SPACE BETWEEN
EYES CONVEX

MODERATE
KNOB
AT END
OF CHIN

BLACK LINING
ON INSIDE OF GUT
CAVITY

POSTERIOR PROFILE OF
ANAL FIN VERTICAL OR
SLANTED FORWARD

SPECIES: *Sebastes ciliatus*—from the Greek *sebastos*, meaning "magnificent;" and the Latin *ciliatus*, "eyelash."

ALTERNATE NAMES: rock cod*.

MAXIMUM RECORDED SIZE: 41 cm (16 inches).

DISTRIBUTION: Cracroft Point, West Cracroft Island on the central coast of British Columbia, to the Bering Sea and perhaps Kamchatka, USSR.

Try spincasting with "buzz bombs," "sting sildas" or jigs, near rocky reefs for the often-schooling dusky rockfish. Light-tackle anglers may also enjoy catching it with bait at depths to 270 m (890 feet).

The active dusky rockfish usually hovers in loosely organized groups just above rocky reefs and along shorelines, or may rest singly upon the rocky substrate. Resembling several other species found in the Pacific Northwest, the dusky rockfish has only recently been detected in British Columbia and knowledgeable divers now commonly recognize it as a distinct fish in the Port McNeil area. Because its dark body "soaks up" the light from a strobe or flash, this rockfish, like some others, requires that underwater photographers adjust their F-stops.

Gulf of Alaska trawlers harvesting between the surface and 30 m (100 feet) net the dusky rockfish along with others over level but solid substrates. However, it forms only a minor part of the annual rockfish take; filleted it sells either fresh or frozen.

A boat passenger sometimes sees dark forms splashing at the surface in pursuit of small schooling silvery shapes that frantically attempt to escape. Although these hungry dark predators may actually be any of several schooling rockfishes—and from such a vantage point, indistinguishable—some may be the dusky one.

Mix dusky rockfish flesh with mashed potatoes, egg, salt and pepper, then form into patties to make "fish cakes" for frying.

*—incorrect

80 Bocaccio

HEAD PROFILE CONCAVE

BODY REDDISH OR OLIVE BROWN

SPACE BETWEEN EYES
CONVEX WITH SPINES
THERE, EITHER
WEAK OR ABSENT

LARGE
PROMINENT
KNOB
AT END
OF CHIN

LARGE MOUTH WITH
GREATLY PROJECTING
LOWER JAW

SPECIES: *Sebastes paucispinus*—from the Greek *sebastos*, meaning "magnificent;" as well as the Latin *pauci* and *spinus*, connoting "few" and "spine."

ALTERNATE NAMES: salmon rockfish, brown bomber, rock salmon*, salmon grouper*, grouper*, rock cod*, tomcod*.

MAXIMUM RECORDED SIZE: 91 cm (36 inches) and 6.8 kg (15 pounds).

DISTRIBUTION: Punta Blanca, central Baja California, Mexico, to Kodiak Island, Gulf of Alaska.

A very popular gamefish in California, the bocaccio, Italian for "large mouth," most often seizes cast or slowly trolled lures, such as herring strip, herring skin or even bucktail flies. Still-fishing with live and tempting herring, anchovy, sea perch or even other small rockfish may also produce catches of this impressive species. While adults dwell primarily in deep water—down to 300 m (1,000 feet)—young bocaccio sometimes hover under floats and provide enjoyment particularly for youthful anglers.

While southern divers often encounter this brown fish, Pacific Northwest aquanauts only occasionally see the bocaccio, in open water adjacent to deeper reefs. Challenge-seeking spearfishing aficionados find the large, free-swimming adult a worthy but often elusive quarry. Look among lush kelp growths along the outer coast and notice the spotted, orange-tinged juveniles hovering there.

Although Pacific Northwest trawlers take some bocaccio, and then sell them along with various other rockfishes, southern Californian commercial interests harvest this species more intensely. Each mature female may perhaps produce a brood twice a year, and a very large female can bear up to 2,300,000 young at one time.

Small juvenile bocaccio, occasionally in large groups, lurk under floats in some locales, but their specific identity is most difficult to verify from above water.

If kept chilled to avoid spoilage, the bocaccio's low-fat flesh is excellent.

*—incorrect

111

81 Silvergray Rockfish

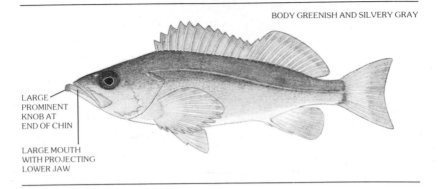

BODY GREENISH AND SILVERY GRAY

LARGE PROMINENT KNOB AT END OF CHIN

LARGE MOUTH WITH PROJECTING LOWER JAW

SPECIES: *Sebastes brevispinus*—from the Greek *sebastos*, meaning "magnificent;" as well as the Latin *brevis* and *spinus*, "short" and "spine."

ALTERNATE NAMES: silvergrey rockfish, shortspine rockfish, short-spined rockfish, short spined rockfish, rock cod *.

MAXIMUM RECORDED SIZE: 71 cm (28 inches).

DISTRIBUTION: Santa Barbara, southern California, to the Bering Sea coast of Alaska.

Although present from the surface to 360 m (1,200 feet) in many locales, anglers seldom catch the large silvergray rockfish. Theoretically it should strike slowly trolled spoons, actively jerked jigs, or still-fished baits, then battle strenuously to escape.

The drab silvergray rockfish usually glides slowly about just above the sea floor but occasionally may retreat into the rocky caverns and crevices of solid formations. Smaller specimens live in shallower depths and the average diver is not likely to swim deep enough to observe the largest ones. After carrying many thousands of minute, poorly developed young, each pregnant silvergray rockfish releases them in late spring. Drifting with the currents, most of these hapless larvae will fall prey to countless other planktonic creatures and only a few young silvergray rockfish will live and settle to the habitat of the grayish-tan adults.

Filleted and sold indiscriminately with other "rockfish" flesh, the increasingly popular silvergray rockfish forms a significant part of the trawl catch taken offshore.

Primarily a deep water denizen, the large-mouth silvergray rockfish seldom invades shallows accessible to surface-viewing naturalists.

Subtle differences in flavour and texture occur between the various rockfishes, with the silvergray's flesh being particularly good. Open your favourite seafood cookbook, turn to the section on white fish, and try this species in any recipe. You'll definitely be pleasantly surprised.

*—incorrect

82 Widow Rockfish

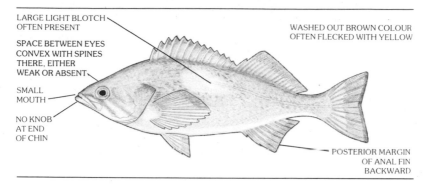

LARGE LIGHT BLOTCH OFTEN PRESENT

SPACE BETWEEN EYES CONVEX WITH SPINES THERE, EITHER WEAK OR ABSENT

SMALL MOUTH

NO KNOB AT END OF CHIN

WASHED OUT BROWN COLOUR OFTEN FLECKED WITH YELLOW

POSTERIOR MARGIN OF ANAL FIN BACKWARD

SPECIES: *Sebastes entomelas*—from the Greek *sebastos*, *ento* and *melas*; meaning "magnificent," "within" and "black."

ALTERNATE NAMES: soft brown, brown bomber, widow rockcod*, rock cod*, bass*.

MAXIMUM RECORDED SIZE: 53 cm (21 inches).

DISTRIBUTION: Bahia Todos Santos, northern Baja California, Mexico, to Kodiak Island, Gulf of Alaska.

A fine light-tackle gamefish, the sporty widow rockfish may often seize "buzz bombs" or other lures at the surface, then battle admirably to escape. At other times, perhaps while angling for salmon, a fisherman may catch one in deeper water.

Scan the greenish midwater above a rocky reef or steep shoreline dive site and perhaps spot the slowly cruising widow rockfish, the brownish one often with a large light blotch on each side. Only occasionally huddling motionless among caves or crevices, the active and free-swimming widow rockfish provides a challenge to even the most proficient spearfishing enthusiast. A female may mature by the end of her third or fourth year. A large female may mate and then, during late winter, jettison as many as 900,000 minute, transparent, living young.

Trawl nets towed along hard and essentially flat bottoms at depths to 370 m (1,220 feet) harvest limited quantities of the saleable widow rockfish, later to be filleted, chilled, then marketed. The disgruntled salmon troller, however, usually discards any widow rockfish swift enough to seize his hardware.

After noticing a splashing and commotion at the surface, then cruising quickly to the immediate area, a boating naturalist may sometimes actually witness the light brown widow rockfish chasing herring or other schooling fishes.

Although reputed to be tougher and coarser than that of other rockfishes, the slender widow rockfish's flesh provides a worthy meal.

*—incorrect

113

83 Canary Rockfish

BRIGHT ORANGE COLOUR

DARK BLOTCH IN JUVENILES

SMOOTH ON UNDER SIDE OF LOWER JAW

3 STRIPES

WHITISH AREA

VERY POINTED

SPECIES: *Sebastes pinniger*—from the Greek *sebastos*, meaning "magnificent;" as well as the Latin *pinna* and *gero*, connoting "fin" and "to bear."

ALTERNATE NAMES: orange rockfish, orange rock-fish, fantail rockfish, fantail, yellow snapper*, red snapper*, red rock cod*.

MAXIMUM RECORDED SIZE: 76 cm (30 inches).

DISTRIBUTION: Cabo Colnett, northern Baja California, Mexico, to Cape Bartolome, southeastern Alaska.

A beautiful and worthy prize, the large, active canary rockfish readily strikes at slow-moving artificial lures or jigs, as well as baits, particularly small live fishes. While a deep-water specimen provides little sport because its gas-filled swim bladder expands and disables the fish as it is hauled up, a shallow-water individual often provides fine light-tackle action. To their chagrin, salmon moochers also sometimes catch canary rockfish.

Watch for the bright-orange canary rockfish hovering in large, loosely organized groups above the rocky bottom. A large specimen is usually in deep water while a juvenile, with a distinct black blotch on the middle of its dorsal fin, swims at shallow, more accessible depths. While stalking the active canary rockfish for different reasons, camera toters and spear carriers both may take up the challenge of "shooting" it.

By dragging their large nets over hard, uneven substrates at depths to 360 m (1,190 feet), trawlers actively harvest the plentiful canary rockfish. Filleted before marketing, it annually ranks in numbers among the top three rockfish species caught. Longliners and salmon trollers may encounter the canary rockfish too, but the latter most likely will discard it.

Very tiny, barely recognizable young canary rockfish, with their black dorsal spots, only rarely hide within tidepools or huddle under floats.

Highly prized, the large canary rockfish delivers a good quantity of delicious, firm flesh.

*—incorrect

114

84 Vermilion Rockfish

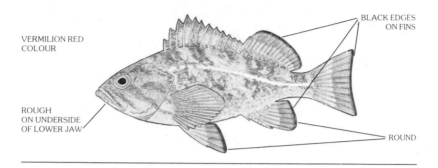

VERMILION RED
COLOUR

BLACK EDGES
ON FINS

ROUGH
ON UNDERSIDE
OF LOWER JAW

ROUND

SPECIES: *Sebastes miniatus*—from the Greek *sebastos* and *miniatus*, meaning "magnificent" and "vermilion."

ALTERNATE NAMES: vermilion rock-fish, rasher, genuine red, red rock cod*, red snapper*.

MAXIMUM RECORDED SIZE: 91 cm (36 inches) and 6.8 kg (15 pounds).

DISTRIBUTION: Isla San Benito and central Baja California, Mexico, to Gillen Harbour, northern British Columbia.

Most popular with anglers in the southern part of its range the vermilion rockfish takes baits such as octopus, squid, herring, anchovies and other small silvery fishes. Fish these baits over rocky reefs, along steep sloping shorelines, or even from wharves to catch this species, which may put up a good battle. A large adult female may be particularly prolific and capable of releasing as many as 1,600,000 tiny young, usually in winter.

Cruising in small schools, pairs or singly below 15 m (50 feet), bright red adult vermilion rockfish are an outstanding and uncommon sight for the sport diver. Underwater photographers particularly enjoy this elusive and shy species that requires patient stalking for that "special shot." A knowledgeable dive buddy can be an effective "herder." Look for the usually more drab, brownish juvenile vermilion rockfish in shallower waters where it may retreat among kelp beds or hole up in rocky crevices.

While only occasionally found in Pacific Northwest trawl catches, the impressive vermilion rockfish ranks third among rockfish species netted in California. There, catches occur in water less than 270 m (890 feet) and the filleted product reaches the market with other "rockfish."

Only very occasionally might a young vermilion rockfish swim beneath a float or around a pier.

Dice mushrooms into vermilion rockfish fillets, rub through a seive, then add bread crumbs and milk preheated in a double boiler; pour together with beaten eggs into custard cups and bake.

*—incorrect

85 Tiger Rockfish

SPACE BETWEEN EYES
STRONGLY CONCAVE

NO SPINES
ABOVE EYE
SOCKETS

PINK OR RED BODY
WITH 5 DARKER BANDS

SPECIES: _Sebastes nigrocinctus_—from the Greek *sebastos*, meaning "magnificent;" and the Latin *niger* and *cinctus*, "black" and "belt."

ALTERNATE NAMES: blackbanded rockfish, banded rockfish, red rock cod*, rockcod*.

MAXIMUM RECORDED SIZE: 61 cm (24 inches).

DISTRIBUTION: Point Buchon, central California, to Cape Ressurection, Kenai Peninsula, Gulf of Alaska.

Boating anglers are the ones who most often catch the darkly banded, pink or red tiger rockfish, because it primarily resides among rocky reefs at depths greater than 10 m (33 feet). Small live fish or silvery jigs seem particularly attractive to this solitary animal which, seldom caught regularly or abundantly, puts up only minimal resistance.

Never straying far from its secure crevice or cavern, the territorial tiger rockfish often erects its large spiny fins and boldly confronts all intruders—including divers. Search for this colourful rockfish on deeper sorties because it lives mainly at depths between 10 and 275 m (33 to 900 feet). Although the tiger rockfish is an inviting spearfishing target, aquanauts most often stalk it with underwater camera gear.

Commercial line fishermen take the tiger rockfish infrequently and incidentally market it as "rockfish." The trawl, seine, or gillnet is totally ineffective for harvesting this inhabitant of deep caverns.

Surface-bound naturalists seldom encounter the adult tiger rockfish, but may easily find juveniles. In spring and summer, look for the striped and kelp-coloured young among the detached seaweeds that accumulate in long tidelines. As tiny tenants in this temporary and drifting habitat, these nomadic juveniles eventually float to rocky areas and descend to adult territory. Juveniles of other rockfishes also huddle beneath flotsam and make specific identification somewhat difficult.

Firm and flavourful. Ideal for any recipe requiring fillets of white-fleshed fish.

*—incorrect

86 Yelloweye Rockfish

STRONG, HIGH AND ROUGH
RIDGES RUNNING PARALLEL
AND BEHIND EACH EYE

SPINE PRESENT
ABOVE
EACH EYE
SOCKET

EYE BRIGHT
YELLOW

SPECIES: *Sebastes ruberrimus*—from the Greek *sebastos*, meaning "magnificent;" and the Latin *ruberrimus*, "very red."

ALTERNATE NAMES: rasphead rockfish, red rockfish, turkey red rockfish, goldeneye rockfish, pot belly, red snapper**, red rock cod*, red cod*, drum*.

MAXIMUM RECORDED SIZE: 91 cm (36 inches).

DISTRIBUTION: Ensenada, northern Baja California, Mexico, to the Gulf of Alaska.

Even though it provides poor sport—its gas-filled swim bladder expands rapidly, often forcing its gut out of its mouth and bulging its eyes as the fish moves toward the surface—seafood-eating anglers pursue the large yelloweye rockfish zealously. Use either jigs or silvery bait-fish, and fish over submerged rocky reefs, particularly where a depth sounder indicates a dropoff, at depths below 45 m (150 feet).

Showing two narrow, bold white stripes along each side, the juvenile yelloweye rockfish is more observable than the large, yellowish-orange adult because the young inhabit shallower regions of a rocky reef or cliff face. While the avid underwater camera buff stalks and films the photogenic juvenile, which often peers out of a deep crevice, the dauntless spear-carrier must usually descend a little deeper to bag the large, slow-swimming and yellow-eyed adult. Look for the popular yelloweye rockfish, adult or juvenile, below 14 m (50 feet).

Although set-lining halibut fishermen once scorned and discarded the now-valued yelloweye rockfish, today they retain it for a good price.

Only in aquaria will the surface-bound naturalist see a live specimen of the popular yelloweye rockfish.

With a lower oil content than that of other rockfishes, the yelloweye rockfish is particularly delicious deep-fried as "fish n' chips."

*—incorrect
**—very popular but incorrect—this is a rockfish and not a member of the snapper family.

117

87 Splitnose Rockfish

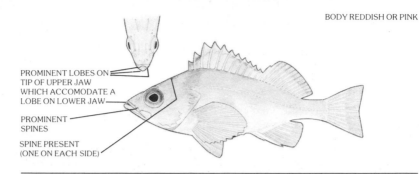

BODY REDDISH OR PINK

PROMINENT LOBES ON
TIP OF UPPER JAW
WHICH ACCOMODATE A
LOBE ON LOWER JAW

PROMINENT
SPINES

SPINE PRESENT
(ONE ON EACH SIDE)

SPECIES: *Sebastes diploproa*—from the Greek *sebastos*, *diploos* and *prora*; meaning "magnificent," "double" and "prow."

ALTERNATE NAMES: lobe-jawed rockfish, small red rock cod*, red snapper*.

MAXIMUM RECORDED SIZE: 46 cm (18 inches).

DISTRIBUTION: Los Coronados Islands, northern Baja California, Mexico, to Prince William Sound, Gulf of Alaska.

Very few anglers catch the large-eyed splitnose rockfish because it lives at great depths from 91 to 578 m (300 to 1,907 feet) and over soft, level and seldom-fished bottoms.

Although abundant and thriving much deeper than divers dare venture, the bright red adult splitnose rockfish only very rarely moves into depths shallower than 36 m (120 feet).

Swept indiscriminately into gaping trawl nets along with other larger and more valuable fish, the moderately-sized splitnose rockfish forms a minor part of the total rockfish catch taken between 200 and 578 m (660 and 1,907 feet). A few of the biggest specimens reach the market as "rockfish" fillets. A large female splitnose rockfish, usually longer than a male of the same age, may give birth to as many as 255,000 young in spring.

Summer boat passengers may find tiny juvenile splitnose rockfish nestling among floating debris dotting long tidelines, or accumulating under floats. Look especially in detached marine plants for this tiny kelp-coloured juvenile which is distinguishable by its large eyes. The most common juvenile "stowing away" among the Georgia Strait tidelines kelp, the young splitnose rockfish at a size of less than 5 cm descends to deep, dark adult habitat by autumn.

When filleting a large splitnose rockfish, do not puncture the internal rubbery pouches located near the anus; a smelly black liquid may ruin the small, short but very sweet-tasting steaks.

*—incorrect

88 Puget Sound Rockfish

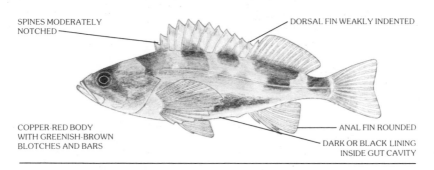

SPINES MODERATELY NOTCHED

DORSAL FIN WEAKLY INDENTED

COPPER-RED BODY WITH GREENISH-BROWN BLOTCHES AND BARS

ANAL FIN ROUNDED

DARK OR BLACK LINING INSIDE GUT CAVITY

SPECIES: *Sebastes emphaeus*—from the Greek words *sebastos*, and *emphaeus*; meaning "magnificent" and "display."

ALTERNATE NAMES: rock cod*.

MAXIMUM RECORDED SIZE: 18 cm (7 inches).

DISTRIBUTION: Northern California to the Kenai Peninsula, Gulf of Alaska.

While after larger reef dwelling bottomfish, the bait-fishing or jigging angler may incidentally catch a Puget Sound rockfish. Do not despair at its small size but instead affix it securely on a large hook, lower it back to the bottom and wait for a sudden, strong tug from another large rockfish, lingcod, or cabezon.

During exploration of reefs or rocky shorelines, divers frequently encounter loose congregations of the usually orange-tan Puget Sound rockfish hovering in open water just slightly above the bottom or nestling warily among crevices and caves. They hole up often during winter or when strong currents are running. Swelled with thousands of minute, developing young, mature female Puget Sound rockfish—easily distinguished from the slimmer males during August and September—will each release as many as 57,000 young. Before the advent of SCUBA, the Puget Sound rockfish remained largely unknown, and still much of its biology and ecology awaits investigation.

Not sought by commercial harvesters, the variably-patterned Puget Sound rockfish rarely shows up in catches as an incidental because a precipitous rocky habitat protects it from most nets, and its small mouth cannot easily seize the large baited hooks of a set-line. Seldom does it enter prawn traps either.

The common Puget Sound rockfish may venture into shallows accessible to beachcombers and wharf-bound naturalists, and is sometimes visible to boat passengers.

Though tasty and flaky, the flesh of the small Puget Sound rockfish rarely graces the dining table.

*—incorrect

89 Redstripe Rockfish

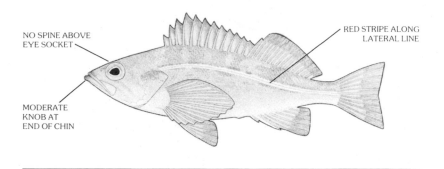

NO SPINE ABOVE
EYE SOCKET

RED STRIPE ALONG
LATERAL LINE

MODERATE
KNOB AT
END OF CHIN

SPECIES: *Sebastes proriger*—from the Greek words *sebastos*, meaning "magnificent;" plus the Latin *prora* and *gero*, connoting "bow" and "to bear."

ALTERNATE NAMES: red-striped rock-fish, rock cod*.

MAXIMUM RECORDED SIZE: 51 cm (20 inches).

DISTRIBUTION: San Diego, southern California, to the Bering Sea coast of Alaska.

 Boaters "mooching" for salmon or bait-fishing near the bottom at depths below 25 m (85 feet) once in a while hook the slender redstripe rockfish incidentally. It seldom puts up much of a scrap even on light tackle and is not highly regarded.

 Look for the pinkish redstripe rockfish, with a lateral red band on each side, at depths in excess of 25 m (85 feet) where rocky reefs and steep silt-covered cliff faces meet gently sloping sandy or muddy bottoms. Night divers may more frequently notice this moderate-sized rockfish as it hovers just off the sea floor or rests upon a rocky perch. A good underwater photograph may be difficult to obtain because this elusive creature often remains just beyond good close-up range.

 Trawlers dragging their bag-like nets along the coast from Oregon to Alaska at depths to 365 m (1,200 feet) frequently take large numbers of redstripe rockfish. After filleting it, commercial processors sell it either fresh or frozen along with other rockfish species.

 A deep water denizen, the redstripe rockfish never swims in water shallow enough to allow the surface-bound naturalist to observe it.

 Of delicate flavour and flaky texture, redstripe rockfish flesh is ideal for poaching in white wine, then garnishing with butter, lemon and parsley. Or why not slice the fresh raw fillets into long narrow strips and dip into soya sauce before eating, Japanese-style.

*—incorrect

90 Greenstripe Rockfish

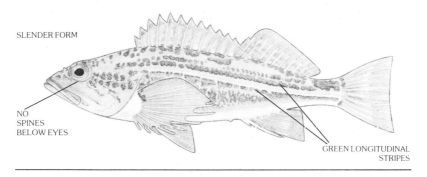

SLENDER FORM

NO
SPINES
BELOW EYES

GREEN LONGITUDINAL
STRIPES

SPECIES: *Sebastes elongatus*—from the Greek words *sebastos*, meaning "magnificent;" plus the Latin *elongatus*, "elongate."

ALTERNATE NAMES: greenstriped rockfish, striped rockfish, strawberry rockfish, poinsetta, strawberry rockcod *.

MAXIMUM RECORDED SIZE: 38 cm (15 inches).

DISTRIBUTION: Isla Cedros, central Baja California, Mexico, to Green Island, Gulf of Alaska.

A smallish fish living deep over level, mostly sandy bottoms, the slender greenstripe rockfish seldom sees any hooked offering because few anglers fish such locales. If available however, this rockfish, with its indistinct green striping, is itself an excellent bait for large lingcod.

Seldom invading water less than 36 m (120 feet) deep, the reddish-tinged greenstripe rockfish rarely encounters sport divers. One of the few rockfish species that consistently lives upon sandy or silty sea floors, it may also cruise near and seek out loose boulders and rocky outcroppings for shelter. A night diver might more readily find the seldom-seen greenstripe rockfish where steep sloping rock formations meet gently sloping sandy substrates.

Commercial trawlers take vast quantities of greenstripe rockfish as part of the total groundfish catch harvested over level, rock-sprinkled bottoms at depths to 402 m (1,320 feet). Only the largest specimens, though, are suitable for processing into fillets; the smaller ones become either animal feed or fish meal. Salmon trollers occasionally catch this fish too.

Except for very rare appearances in public aquariums, the bottom-dwelling greenstripe rockfish swims at depths completely out of view to the surface-bound naturalist.

While the greenstripe rockfish is somewhat small, its flesh has a distinct delicious flavour and an appealing texture. After filleting, coat the tiny but worthwhile steaks with bread crumbs, then fry briefly.

* —incorrect

91 Shortspine Thornyhead

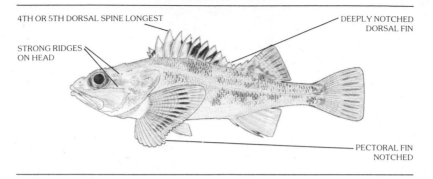

4TH OR 5TH DORSAL SPINE LONGEST

DEEPLY NOTCHED
DORSAL FIN

STRONG RIDGES
ON HEAD

PECTORAL FIN
NOTCHED

SPECIES: *Sebastolobus alascanus*—from the Greek *sebastos* and *lobos*, meaning "magnificent" and "lobe;" and emphasizing Alaska.

ALTERNATE NAMES: spinycheek rockfish, spiny-cheek rockfish, shortspine channel rockfish, spinyheaded rockfish, lobe-finned rockfish, thornyhead, idiotfish, scorpion, bonehead, idiot, hooligan, red rock cod*, channel rockcod*, channel cod*, gurnard*, gurnet*.

MAXIMUM RECORDED SIZE: 75 cm (29.4 inches).

DISTRIBUTION: northern Baja California, Mexico, to the Bering Sea and the Okhotsk Sea, USSR.

Very few bait-fishermen lower tackle to depths greater than 80 m (264 feet) where the shortspine thornyhead flourishes.

Because the bright-red shortspine thornyhead has not yet been recorded in water less than 80 m (264 feet), a SCUBA sighting of one would indeed be noteworthy. As yet only passengers aboard submersibles view this creature resting upon its favourite sandy or silty bottom.

The colourful shortspine thornyhead lives at depths to 1,524 m (5,000 feet) where it is often taken incidentally by groundfish and shrimp trawlers, or occasionally on sablefish longlines. While this slender species seldom reaches the fish market, many professional fishermen sort it from the catch for their own use.

Although the deep water shortspine thornyhead is obviously inaccessible to surface-bound naturalists, its drifting and fertilized egg masses are not. Always found at night and floating at the surface, variously shaped hollow gelatinious egg balloons up to 60 cm (24 inches) long, each encase a single layer of many tiny eggs. Within ten days of breaking free from the egg balloon, tiny larvae at first derive nourishment from their yolk sacs then later begin feeding upon their minute planktonic neighbours.

The shortspine thornyhead's white flesh is somewhat sweeter than that of most other rockfish.

*—incorrect

122

THE SABLEFISHES
(Family: Anoplopomatidae)

Related to the rockfishes, sculpins, greenlings, poachers and snailfishes through having suborbital stays—a small, heavy bone beneath each eye—the very tiny family Anoplopomatidae has existed in the waters of this planet for at least 12 million years, since the Miocene epoch. Only two living species are known: one is the popular and common sablefish; and the other, the large skilfish, an offshore Pacific Northwest creature rarely seen at sea by naturalists.

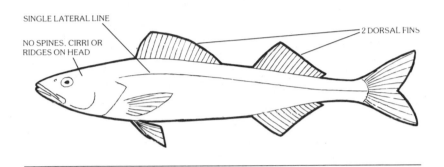

SINGLE LATERAL LINE

NO SPINES. CIRRI OR
RIDGES ON HEAD

2 DORSAL FINS

John Lozanski, a meterological technician for Environment Canada, is one person fortunate enough not only to have seen several live skilfish, but also to have captured and transported them alive to a public aquarium for other people to enjoy. During a mid-summer cruise of the Canadian weathership *Vancouver*, while fishing for salmon, Mr. Lozanski was surprised to sight six stout juvenile skilfish, each about 30 cm (l foot) long. The dark blue shapes boldly blotched with large white patches, had pursued his tackle to the surface but seemed unafraid, and instead of quickly fleeing they inquisitively milled about beside the immense vessel; Mr. Lozanski even had time to run to the stern of the ship, grab a longhandled dipnet, return, then deftly capture these curious creatures. Within an hour of being placed in a specially provided aquarium tank located in the laboratory of the ship, the friendly captives were feeding from his hand, a behaviour most unusual for "wild" and just-captured fish. Growing half again their initial capture size during the remainder of the cruise, the exceedingly tame skilfish, now the ship's pets, were eventually transported to larger permanent quarters at the Vancouver Public Aquarium. For nearly fifteen years, these amazing specimens delighted visitors and staff alike, growing very large but well short of their potential dimensions of 91 kg (200 pounds) and 178 cm (70 inches).

92 Sablefish

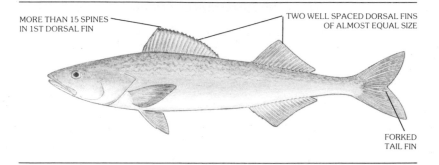

MORE THAN 15 SPINES
IN 1ST DORSAL FIN

TWO WELL SPACED DORSAL FINS
OF ALMOST EQUAL SIZE

FORKED
TAIL FIN

SPECIES: *Anoplopoma fimbria*—from the Greek *anoplis* and *poma*, meaning "unarmed" and "gill cover;" plus the Latin *fimbria*, "fringed."

ALTERNATE NAMES: Alaska blackcod, blackcod, coalfish, coalcod, black cod, blue cod*, bluefish*, candlefish*, skil-fish*, skil*.

MAXIMUM RECORDED SIZE: 107 cm (42 inches) and over 57 kg (126 pounds).

DISTRIBUTION: Isla Cedros, central Baja California, Mexico, to the Bering Sea coasts of Alaska, and Kamchatka, USSR, and south to Hatsu Shima Island, southern Japan.

Abundant enough to be a nuisance in some areas, young sablefish often pester salmon anglers who troll or mooch with herring. On the other hand, pier-bound bait-fishermen enjoy the light-tackle sport the active juvenile sablefish provide when swarming around pilings, jetties, and wharves during summer.

Divers exploring the sandy, silty bottoms of shallow bays might possibly see the sleek gray shapes of young sablefish cruising near the bottom. Do not expect to come very close to this fish, though, because it is wary and swift swimming.

Handliners, setliners, trawlers and even fishermen using converted crab traps have all harvested sablefish over the years, from the surface to depths of 1,829 m (6,000 feet). Traditionally the native Indians sun-dried the oily sablefish but now commercial fishermen prize it highly and market it extensively as "smoked Alaska black cod."

That the surface-bound observer will sight a sablefish, even in shallow and clear water around wharves, is unlikely.

Using either a home-made or an inexpensive retail smoker, anyone may create their own delicious smoked Alaska black cod. Do a little reading and some experimenting. Smoke only large, adult sablefish, though, because their high oil content ensures success. The orange or yellow colour of the commercially available product is an additive and unnecessary.

* —incorrect

THE GREENLINGS
(Family: Hexagrammidae)

Greenlings and combfishes, along with the lingcod and Atka mackerel, make up the Hexagrammidae, a small family of fishes containing only 13 known living species; about half of these reside along the west coast of North America and the rest in the Asian Pacific. Except for the voracious lingcod, the largest of the clan at 152 cm (5 feet), most greenlings prefer shallow, even intertidal, rocky or weedy bottoms. Here these colourful aggressive fish stake out small but definite territories and staunchly defend them, particularly during mating season.

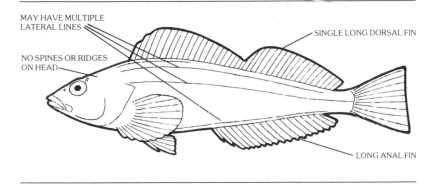

MAY HAVE MULTIPLE LATERAL LINES

NO SPINES OR RIDGES ON HEAD

SINGLE LONG DORSAL FIN

LONG ANAL FIN

With only a limited amount of known fossil material to study, paleontologists believe ancestral hexagrammids populated the earth's shallow seas as long ago as 25 million years, in the Oligocene period. Together with the many sculpins, rockfishes, sablefishes, poachers, and snailfishes, the common greenlings have heavy bones called suborbital stays, located under their eyes, and form the large category called mail-checked fishes. With the notable exception of the large and valuable lingcod, the Pacific Northwest greenlings remain relatively unexploited because modern trawl techniques do not function well over uneven, shallow substrates. Perhaps, though, future world food needs will stimulate development of other harvesting methods to exploit the very edible hexagrammids.

Several Pacific Northwest greenlings not commonly observed are the masked greenling, a northern species with 4 lateral lines on each side, the Atka mackerel with its deeply forked tail, and the longspine combfish, noteworthy for the very long first two spines on its dorsal fin.

93 Kelp Greenling

2 PAIRS OF SMALL THREAD-LIKE PROJECTIONS BEHIND EYES

(FEMALE)

LIGHT MARK DARKLY MARGINED

SMALL SINGLE SPINE (BOTH SEXES)

(MALE)

SPECIES: *Hexagrammos decagrammus*—from the Greek *hexa*, *deca* and *gramma*, meaning "six," "ten" and "line."

ALTERNATE NAMES: greenling sea trout*, speckled sea trout*, rock trout*, bluefish*, tommy cod*.

MAXIMUM RECORDED SIZE: 61 cm (24 inches) and 2.l kg (4.6 pounds)

DISTRIBUTION: Santa Monica Bay, southern California, to Amchitka Island, in the Aleutian chain, Alaska.

 Because the common kelp greenling bites at nearly any bait, it is very popular with bait-fishermen who cast into shallow water from the shore, from a jetty or even from a boat. Jerk a silvery jig constantly and easily to catch this abundant fish, itself an excellent bait for its large relative, the lingcod. Use light tackle to enjoy the scrappy, active kelp greenling most.

 A very common companion of divers in the Pacific Northwest, the inquisitive kelp greenling often follows the aquanaut and often accepts hand-held food—a valuable aid for the underwater photographer. Once believed to be two different species, the adult male and adult female both flourish at depths less than 15 m (50 feet), where they lurk among plant life or boulders. During autumn and winter, adult male kelp greenling—usually very pale in coloration—actively guard clusters of mauve or pale blue coloured eggs deposited earlier by one or more females.

 Only occasionally taken by commercial line fishermen, and even less often by trawlers, the potentially valuable kelp greenling rarely reaches market.

 While strolling along rocky shores or near tidepools, especially in summer, the observant beachcomber may notice the colourful kelp greenling, particularly the silvery young which are less than 5 cm long. The Aleut Indians prize this species and call it *idyajuk*.

 Try filleting the tasty kelp greenling, then cooking it with yogurt and some favourite spices for a low-calorie seafood meal.

*—incorrect

94 Whitespotted Greenling

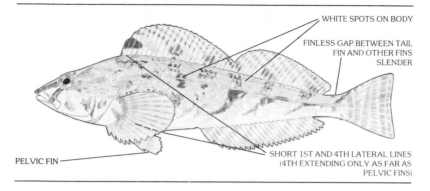

WHITE SPOTS ON BODY

FINLESS GAP BETWEEN TAIL
FIN AND OTHER FINS
SLENDER

PELVIC FIN

SHORT 1ST AND 4TH LATERAL LINES
(4TH EXTENDING ONLY AS FAR AS
PELVIC FINS)

SPECIES: *Hexagrammos stelleri*—from the Greek *hexa*, and *gramma*, meaning "six" and "line;" and honouring George W. Steller, the famous Russian naturalist.

ALTERNATE NAMES: rock trout*, tommy cod*, kelp cod*.

MAXIMUM RECORDED SIZE: 48 cm (19 inches) and 1.6 kg (3.6 pounds).

DISTRIBUTION: Coos Bay, southern Oregon, to the Aleutian Island chain, Bering Sea, Peter the Great Bay, USSR, and Hokkaido, northern Japan.

Youthful anglers, especially, find the easily caught whitespotted greenling an enjoyable summertime quarry. This olive-hued fish snaps particularly at natural baits such as marine worms, small crabs and fishes when bottom-fished from jetties and wharves. Small and not too scrappy.

The active whitespotted greenling lives on shallow sandy bottoms, mainly among eelgrass or other marine plants growing near rocky outcroppings at depths less than 15 m (50 feet). In winter, the golden males with their temporarily darkened fins fearlessly guard the egg masses deposited by females in cracks of rocky outcroppings and may even boldly swim circles around divers. After hatching from small pale blue eggs, the tiny larvae feed upon minute young crabs, barnacles and floating fish eggs. By mid-spring, though, the young whitespotted greenlings have grown to about 5 cm and have passed through a silvery, surface-dwelling stage before settling to a bottom residence.

An unsought, unmarketed species, the shallow water whitespotted greenling is very seldom a commercial trawl catch.

While not usually trapped in tidepools, the common whitespotted greenling lives in shallows, often among eelgrass, and may be visible to an observant wharf-bound naturalist.

Toast some bread; on one side place boiled whitespotted greenling, cheese, catsup and then broil for tasty canapes.

*—incorrect

127

95 Rock Greenling

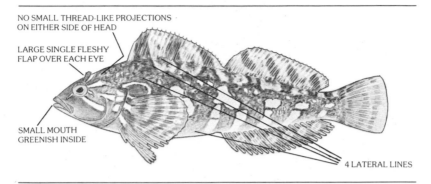

NO SMALL THREAD-LIKE PROJECTIONS
ON EITHER SIDE OF HEAD

LARGE SINGLE FLESHY
FLAP OVER EACH EYE

SMALL MOUTH
GREENISH INSIDE

4 LATERAL LINES

SPECIES: *Hexagrammos lagocephalus*—from the Greek *hex*, *gramma*, *lagos* and *cephalus*, meaning "six," "line," "hare" and "head."

ALTERNATE NAMES: fringed greenling, red rock trout*.

MAXIMUM RECORDED SIZE: 61 cm (24 inches).

DISTRIBUTION: Point Conception, central California, to the Bering Sea and the Kurile Islands, USSR.

A worthy but somewhat elusive species, the vivid rock greenling lives among open coast seaweeds along rugged, surf-pounded, rocky shorelines at depths less than 16 m (55 feet). If adventurous enough to fish such locales, use baits such as clams, small crabs and marine worms, but take along rain gear for protection against salt spray or heavy rain. After seizing a hook, this active greenling usually retreats into crevices—try to make this fish head shoreward immediately.

Both camera buffs and spearfishermen find the adult rock greenling a challenging quarry. If not discouraged by the heavy surge often present along outer rocky shores, look for this turquoise, red and white species; although it should be almost garishly obvious, it is well camouflaged among the surrounding plants. In summer, small juvenile rock greenling, distinguished by their bright red eyes, may nestle here too, or live in more sheltered nearby bays.

Few commercial fishermen venture into the non-schooling rock greenling's forbidding habitat to harvest anything.

Very difficult to see among the colourful plant life growing luxuriantly along the outer rocky coast, the gaudy rock greenling may actually rest within arm's length of even the most diligent beachcomber. A strong dip net scooped through the weeds is the only way to gather an unseen specimen from a tidepool or from nearby shorelines.

Simmer flour, cooked onion, pepper, tomato, brown sugar, vinegar and mustard; then add rock greenling strips with kidney beans and simmer again.

*—incorrect

128

96 Lingcod

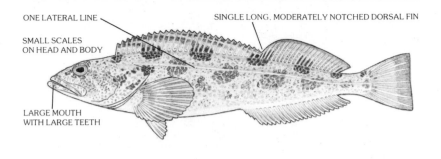

ONE LATERAL LINE

SINGLE LONG, MODERATELY NOTCHED DORSAL FIN

SMALL SCALES
ON HEAD AND BODY

LARGE MOUTH
WITH LARGE TEETH

SPECIES: *Ophiodon elongatus*—from the Greek *ophis* and *odons*, meaning "snake" and "tooth;" and the Latin *elongatus*, "elongate."

ALTERNATE NAMES: Pacific cultus, ling *, buffalo cod *, blue cod *, cultus cod *.

MAXIMUM RECORDED SIZE: 152 cm (60 inches) and 45 kg (100 pounds).

DISTRIBUTION: Ensenada, northern Baja California, Mexico, to the Shumagin Islands, near the Alaskan Peninsula.

The most important non-salmonid gamefish in the Pacific Northwest, the aggressive lingcod lives primarily over rocky bottoms at depths to 2,000 m (6,600 feet). Be sure to observe local bag-limits and seasonal restrictions. Fish for this ravenous species along shores and over reefs, particularly where strong currents flow, and use heavy silvery jigs or small live fish as bait. A powerful predator that may seize other small just-hooked gamefish, the scrappy lingcod usually fights hardest at the surface. Use a sturdy rod and strong line.

The prize target for Pacific Northwest spearfishermen, the large lingcod generally remains within its established territory. Be certain your weapon is adequately powered for the size of the target, lest both spear and wounded lingcod be lost! Actually, moderate-sized specimens are the best eating and the large "trophy" lingcod, always females, should be spared as brood stock. Consult local lingcod bag-limits and seasonal restrictions. Conservation-minded enthusiasts should also leave any male that guards a large white egg mass—even if no restrictions exist.

A vital economic commodity, the popular lingcod fills quotas for longliners, handliners and trawlers before most of it enters the fish n' chip trade.

Clear water occasionally permits the shorebound naturalist to see an adult lingcod; eelgrass often hides juveniles during their first year. While chasing herring near the surface, small specimens may even leap clear of the water.

Lingcod is delicious, even the occasional blue-green fleshed specimen, which becomes white after cooking.

* —incorrect

97 Painted Greenling

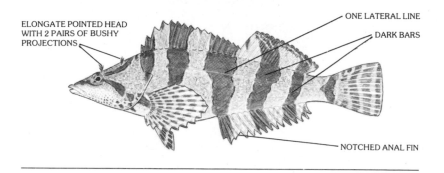

ELONGATE POINTED HEAD WITH 2 PAIRS OF BUSHY PROJECTIONS

ONE LATERAL LINE

DARK BARS

NOTCHED ANAL FIN

SPECIES: *Oxylebius pictus*—from the Greek *oxys* and *lebius*, meaning "sharp" and "kettle;" and the Latin *pictus*, "picture."

ALTERNATE NAMES: convictfish.

MAXIMUM RECORDED SIZE: 25 cm (10 inches).

DISTRIBUTION: Point San Carlos, central Baja California, Mexico, to Kodiak Island, Gulf of Alaska.

In some locales, the painted greenling actively skulks around pilings and under wharves. Although catchable on small hooks baited with piling worms, snails, mussels, shrimps or tiny crabs, this animal directly interests few anglers because it is so small. It might be used, however, as live bait for lingcod or rockfishes.

Active in shallow rocky habitats, the distinctive painted greenling— generally red stripes on a whitish background—lurks singly or in pairs. Seemingly unafraid as it moves from one perch to another, it often trails the diver. When seeking the most colourful specimens, the aquanaut should patrol areas profusely encrusted with a variety of brightly hued invertebrate animals. On rare occasions, really lucky divers may meet the young painted greenling nestling among the large tentacles of sea anemones, apparently immune to the tentacular stinging cells: a great photographic opportunity. Throughout a winter spawning season the mature male is often almost totally black beside the typically striped female, and once a female deposits a cluster of eggs in a rocky crevice, the pugnacious male fertilizes them and then guards the nest.

The economically valueless painted greenling rarely invades prawn traps set on or near rocky bottoms at depths less than 49 m (160 feet). Net fishermen do not set their equipment in such areas.

Provided the water is clear, the dockside observer might notice a painted greenling. Tidepool sightings are improbable because this animal apparently lives deeper than the lowest intertidal pool.

Even the largest painted greenling has only a limited quantity of tasty flesh.

THE SCULPINS
(Family: Cottidae)

The Cottidae, an enormous family, contains approximately 350 described living species including sculpins, Irish lords and sea ravens, as well as the cabezon. An adaptable group, the cottids lurk in both fresh water and marine habitats, while a few also tolerate the intermediate brackish environments. Virtually all depths of water and types of substrate provide habitat for these sluggish creatures, nearly all of which are bottom dwellers as adults. Almost all cottids live in the Arctic and temperate waters of the northern hemisphere, and only a few known species exist south of the equator.

Aside from the Pacific Northwest's cabezon, at 99 cm (39 inches), most cottids grow only to very modest sizes and so few commercial fishermen even attempt to fillet the little marketable flesh from the long tapering bodies. Consequently, only the occasional underwater photographer, aquarium enthusiast, or scientist finds these large headed fishes of any value.

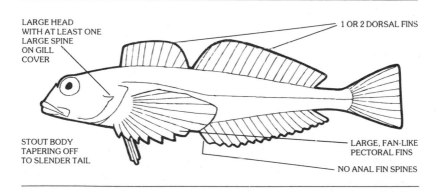

LARGE HEAD WITH AT LEAST ONE LARGE SPINE ON GILL COVER

1 OR 2 DORSAL FINS

STOUT BODY TAPERING OFF TO SLENDER TAIL

LARGE, FAN-LIKE PECTORAL FINS

NO ANAL FIN SPINES

A number of sculpins are seldom seen in the Pacific Northwest. Some of these live at inaccessible depths: the flabby sculpin, the frogmouth sculpin, the fringed sculpin, the thorny sculpin, the smallsail sculpin, the blob sculpin, and the armourhead sculpin. A few other cottids have their centres of abundance in California and Alaska: the bonehead sculpin—similar to the species on page 132 but with no scales below the front of each eye, the coralline sculpin—similar to the species on page 132 but with a small patch of scales just behind each upper gill flap, the bald sculpin—similar to the species on page 140 but with no cirri at the anterior of the space between the eyes, the crested sculpin—similar to the species on page 158 but with no silvery body markings, the leister sculpin—similar to the species on page 165 but with fleshy flaps dangling from the lower head region, and the bigmouth sculpin with its huge mouth.

98 Smoothhead Sculpin

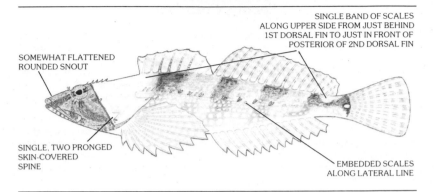

SINGLE BAND OF SCALES
ALONG UPPER SIDE FROM JUST BEHIND
1ST DORSAL FIN TO JUST IN FRONT OF
POSTERIOR OF 2ND DORSAL FIN

SOMEWHAT FLATTENED
ROUNDED SNOUT

SINGLE, TWO PRONGED
SKIN-COVERED
SPINE

EMBEDDED SCALES
ALONG LATERAL LINE

SPECIES: *Artedius lateralis* —honouring the Russian naturalist Petrus Artedi; and the Latin *lateralis*, meaning "on the side."

ALTERNATE NAMES: flathead sculpin, bullhead*.

MAXIMUM RECORDED SIZE: 14 cm (5.5 inches).

DISTRIBUTION: Bahia San Quintin, northern Baja California, Mexico, to the Bering Sea coast of Alaska, along the Aleutian chain to the Komandorski Islands, USSR.

Generally dismissed as a tiny, bait-stealing nuisance by the pier-fishing crowd, the unpopular smoothhead sculpin often grabs any hooked offering and charges off into the pilings but usually loses this meal because the bulky morsel is too big. Resourceful young anglers may use as bait those few specimens actually caught.

One of many small sculpins a diver may encounter, the well-camouflaged smoothhead sculpin lurks at less than 14 m (46 feet); therefore, at the start or finish of a shore dive, look for this species in the shallows. Look closely, though, because it is difficult to locate among weed-covered rocks or upon encrusted vertical pilings. In winter each gravid female smoothhead sculpin lays small clusters of adhesive, cherry-red, yellow, or orange eggs under rocks and within three weeks tiny transparent larvae hatch and swim to the surface. By spring recognizable young smoothhead sculpins settle to the bottom.

The unmarketable smoothhead sculpin lives in shallow and therefore largely unharvested waters, and so is never caught.

In tidepools or under moist rocks left exposed by the receding tide, the variably-hued smoothhead sculpin is difficult to see because its irregular banded coloration blends too well with its background. Once alerted by movement, dockside naturalists may see this inactive creature resting upon pilings or piling supports.

Too small for serious culinary consideration.

*—incorrect

132

99 Padded Sculpin

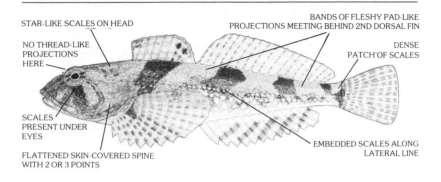

STAR-LIKE SCALES ON HEAD

NO THREAD-LIKE
PROJECTIONS
HERE

BANDS OF FLESHY PAD-LIKE
PROJECTIONS MEETING BEHIND 2ND DORSAL FIN

DENSE
PATCH OF SCALES

SCALES
PRESENT UNDER
EYES

FLATTENED SKIN-COVERED SPINE
WITH 2 OR 3 POINTS

EMBEDDED SCALES ALONG
LATERAL LINE

SPECIES: *Artedius fenestralis*—honouring the Russian naturalist Petrus Artedi; and the Latin *fenestralis*, meaning "with a window."

ALTERNATE NAMES: bullhead*.

MAXIMUM RECORDED SIZE: 14 cm (5.5 inches).

DISTRIBUTION: San Luis Obispo Cove, southern California, to Unalaska Island, and the Alaska Peninsula.

A small, slow-moving creature, the unwelcome padded sculpin primarily represents a bait-thieving annoyance to wharf-bound anglers who use small fishes, shrimps or other items to tempt more desirable quarry. As bait itself, this common creature is of modest value.

Most often sighted by divers exploring the jetty and piling habitat, the cryptically coloured padded sculpin also lurks among eelgrass growing intertidally or slightly deeper, but does live at depths to 60 m (200 feet). Look very carefully to find it, for this usually motionless sculpin is very difficult to spot, and then once disturbed darts headlong for nearby shelter. During winter, gravid female padded sculpins deposit small clusters of purple or gray eggs upon the bottom, under rocks.

A very improbable incidental catch, the unsaleable padded sculpin lives in shallow, commercially unfished locales.

Peering into the water, the dockside observer may notice the gray padded sculpin resting among the barnacles and mussels growing upon the pilings, but this often stationary and irregularly marked creature blends beautifully with any background it has chosen. Although the padded sculpin has a more pointed snout than the smoothhead sculpin, this difference is of modest help to one looking down from above the water. Seaside strollers should look for padded sculpins along the rocky shore, either in tidepools or beneath adjacent rocks.

Too small for serious eating.

*—incorrect

100 Scalyhead Sculpin

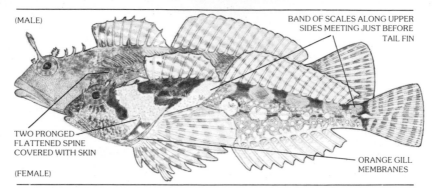

(MALE)

BAND OF SCALES ALONG UPPER SIDES MEETING JUST BEFORE TAIL FIN

TWO PRONGED FLATTENED SPINE COVERED WITH SKIN

(FEMALE)

ORANGE GILL MEMBRANES

SPECIES: *Artedius harringtoni*—honouring two people associated with the Pacific Northwest; Petrus Artedi, early Russian naturalist; and Mark Harrington, past president of the University of Washington.

ALTERNATE NAMES: plumose sculpin, white-spotted sculpin, bullhead *.

MAXIMUM RECORDED SIZE: 10 cm (4 inches).

DISTRIBUTION: San Miguel Island, and southern California, to Kodiak Island, Gulf of Alaska.

An incidental, unwanted but often common catch for bait-fishermen who angle from docks or floats, the pesky scalyhead sculpin is too small to be of much interest.

A very abundant inhabitant of shallow reef or rocky shore locales, the intricately patterned scalyhead sculpin actively darts about amid the colourful attached plants and animals. Look also among the life that encrusts pilings to find this bold sculpin. Photographers find this common colourful little fish a subject that tests their skills with a "macro" lens. Although yellow gill membranes distinguish an adult scalyhead sculpin from most other cottids, juvenile *Artedius harringtoni* are easily confused with other young, particularly those of the Puget Sound sculpin (*Artedius meanyi*), a fish which is heavily scaled above the lateral lines and has a large tri-branched cirrus behind each eye.

The small scalyhead sculpin is not a commercial species but may invade prawn traps, or shrimp trawl nets may drag them up from depths less than 25 m (82 feet).

Tidepools formed at the very lowest tides often strand the common scalyhead sculpin until the returning sea frees it again. However, even a studious beachcomber may have difficulty in spotting one; a fine-mesh dipnet scooped through the weeds may produce a specimen. Dockside observers more readily notice this creature perched upon pilings and their supports.

Too small for serious eating.

* —incorrect

101 Tidepool Sculpin

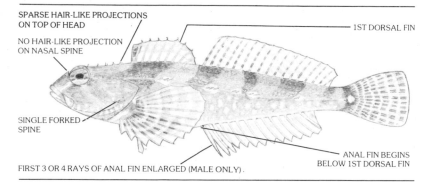

SPARSE HAIR-LIKE PROJECTIONS ON TOP OF HEAD

NO HAIR-LIKE PROJECTION ON NASAL SPINE

1ST DORSAL FIN

SINGLE FORKED SPINE

ANAL FIN BEGINS BELOW 1ST DORSAL FIN

FIRST 3 OR 4 RAYS OF ANAL FIN ENLARGED (MALE ONLY).

SPECIES: *Oligocottus maculosus*—from the Greek *oligo* and *cottus*, meaning "few" and "sculpin;" as well as the Latin *macula*, "spot."

ALTERNATE NAMES: tide pool sculpin, tidepool johnny, bullhead*.

MAXIMUM RECORDED SIZE: 8.9 cm (3.5 inches).

DISTRIBUTION: White Point, near Los Angeles, southern California, to the Bering Sea, through the Aleutian chain to the Kurile Islands and Okhotsk Sea, USSR.

Too small to be caught by the tiniest hook, the abundant tidepool sculpin lives in the intertidal zone where it is easily collected by hand or net for bait.

While gearing up before a shore dive or removing equipment afterwards, notice the active tidepool sculpin darting near your feet and chasing tiny animals you have dislodged. Relax and watch this intriguing performance while waiting for your slower dive buddies to finish with their equipment.

Small enough to slip through virtually all net meshes, the shallow water tidepool sculpin never enters commercial catches.

Certainly the most commonly seen intertidal fish in the Pacific Northwest, the mobile tidepool sculpin frequently darts along the shore with the ebb and flow of the sea, but is also often trapped in even the smallest tidepool or under a rock. Very tolerant of high temperatures and low salinities in the intertidal realm, this variably-coloured sculpin may easily return to its "home" territory when dislocated by man, animal, wave, or tide. A sense of smell apparently guides the homing fish; specimens experimentally blinded by scientists still returned to their original territory. Mating season for this fish occurs from November to May when, after copulation with a well-endowed male, a female deposits small clusters of emerald green or maroon adhesive eggs among mussels or barnacles.

Too small to be of interest.

*—incorrect

102 Fluffy Sculpin

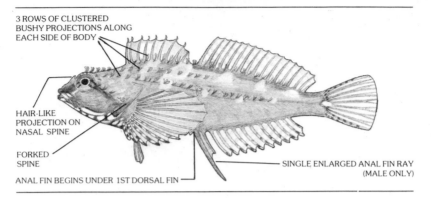

3 ROWS OF CLUSTERED
BUSHY PROJECTIONS ALONG
EACH SIDE OF BODY

HAIR-LIKE
PROJECTION ON
NASAL SPINE

FORKED
SPINE

ANAL FIN BEGINS UNDER 1ST DORSAL FIN

SINGLE ENLARGED ANAL FIN RAY
(MALE ONLY)

SPECIES: *Oligocottus snyderi*—from the Greek *oligo* and *cottus*, meaning "few" and "sculpin;" as well as honouring an early American ichthyologist, John O. Snyder.

ALTERNATE NAMES: cirriated sculpin, bullhead*.

MAXIMUM RECORDED SIZE: 8.5 cm (3.5 inches).

DISTRIBUTION: Rio Socorro, northern Baja California, Mexico, to Sitka, southeastern Alaska.

Perhaps a keen angler might wander out to an exposed coast, don waders, and gather the plentiful fluffy sculpin for bait.

For a change, leave the tank and regulator behind, but take your snorkeling gear to look along exposed rocky-bottomed shallows or in nearby tidepools. Try enticing this reclusive creature into the open with cracked mussels.

Commercial fishermen never venture into the wave-pounded, shallow, rocky shores that are home to the fluffy sculpin.

Capturing fluffy sculpins with a fine-meshed dipnet is often the best way for a close look. Intolerant of higher temperatures and lower salinities of upper intertidal pools, the adult fluffy sculpin lurks in tidepools at lower exposed zones where it hides among the lush plant or encrusting animal life. Some specimens may be intricately marked with pink or red and may nestle among the similarly coloured corralline algae while others, which are bright emerald, huddle amid the identically-hued green surf grass. Also living among these plants are worms and small shrimp-like creatures which are the prey of this active sculpin. During winter, a ripe male will actually clasp a smaller, receptive female by wrapping his prehensile first anal fin ray around her body before copulating. By spring, many tiny young populate exposed tidepools, avoiding established adults by seeking pools higher in the intertidal zone.

Too small for eating.

*—incorrect

136

103 Saddleback Sculpin

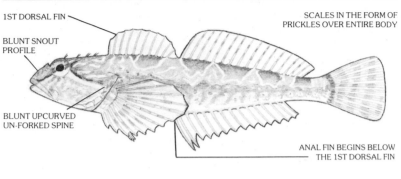

1ST DORSAL FIN

SCALES IN THE FORM OF
PRICKLES OVER ENTIRE BODY

BLUNT SNOUT
PROFILE

BLUNT UPCURVED
UN-FORKED SPINE

ANAL FIN BEGINS BELOW
THE 1ST DORSAL FIN

SPECIES: *Oligocottus rimensis*—from the Greek *oligo*, *cottus*, and *rimensis*, meaning "few," "sculpin," and "swordcleft."

ALTERNATE NAMES: prickly sculpin*, bullhead*.

MAXIMUM RECORDED SIZE: 6.3 cm (2.5 inches).

DISTRIBUTION: St. Nicholas Island and northern Baja California, Mexico, to Kakul Narrows, southeastern Alaska.

The greenish saddleback sculpin never takes an angler's hook.

Only "blue ribbon" fishwatching divers sight the tiny saddleback sculpin because it usually frequents murky shallows along silty or sandy shorelines. Poor visibility, particularly among luxuriant eelgrass growths, obscures this seldom-noticed fish. Because it often perches upon golden brown kelp fronds, look among these plants in slightly deeper, clearer water.

The minute and unmarketable saddleback sculpin lives in shallow or intertidal areas which commercial fishermen usually avoid.

Amid dense growths of marine plants—along the seashore or in tidepools—the obscure saddleback sculpin stakes out its territory. This active fish may also hide among the plants and animals dangling from floats or pilings. Its colour pattern often blends so well with its surroundings that the little creature must move before even the most observant naturalist would see it. Watch it dart about in search of tiny prey or a potential mate while avoiding its own numerous predators such as the great blue heron. To be certain that specimens are not the similar tidepool sculpin, capture some with a dipnet, then inspect them with a magnifying glass. Be a good conservationist—release alive any unwanted creatures.

Too small for human consumption, the diminutive saddleback sculpin may be useful as cat food.

*—incorrect

104 Sharpnose Sculpin

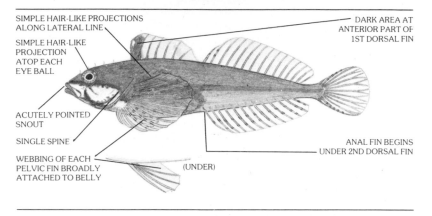

SIMPLE HAIR-LIKE PROJECTIONS ALONG LATERAL LINE

SIMPLE HAIR-LIKE PROJECTION ATOP EACH EYE BALL

ACUTELY POINTED SNOUT

SINGLE SPINE

WEBBING OF EACH PELVIC FIN BROADLY ATTACHED TO BELLY

DARK AREA AT ANTERIOR PART OF 1ST DORSAL FIN

ANAL FIN BEGINS UNDER 2ND DORSAL FIN

(UNDER)

SPECIES: *Clinocottus acuticeps*—from the Greek *clinus* and *cottus*, meaning "blenny" and "sculpin;" as well as the Latin *acutis* and *ceps*, connoting "sharp" and "head."

ALTERNATE NAMES: bullhead˙.

MAXIMUM RECORDED SIZE: 6.3 cm (2.5 inches).

DISTRIBUTION: Big Sur, central California, to the Bering Sea coast of Alaska, and Attu Island in the Aleutian chain.

With a mouth too small to take even the most minute hook, the tiny sharpnose sculpin very seldom becomes a useful bait.

Only observant snorkelers or divers who search diligently along shorelines precisely where the water's surface meets the rocks might spy the active but elusive sharpnose sculpin. Generally, though, in such shallow water, poor visibility associated with wave action makes sighting this tiny creature extremely difficult.

The diminutive sharpnose sculpin lives at the very surface of a rocky environment where commercial enterprises do not harvest.

While strolling along the seashore, look for the colourful sharpnose sculpin, particularly on seaweed covered sea walls or rocks where it follows the ebb and flow of the tide and remains within centimeters of the surface. This fish's colour is most often an olive green like rockweed, but may be golden brown like the kelps, or bright green like sea lettuce. In early spring a swollen female sharpnose sculpin lays small clusters of pale amber adhesive eggs which the smaller adult male fertilizes with his very large genital papilla. Able to tolerate the lower salinities brought about by heavy rains, this species, particularly when young, may also venture into the brackish water of river estuaries. A fine-meshed dipnet is helpful in capturing this hardy and tiny sculpin.

Too small for eating.

˙—incorrect

105 Calico Sculpin

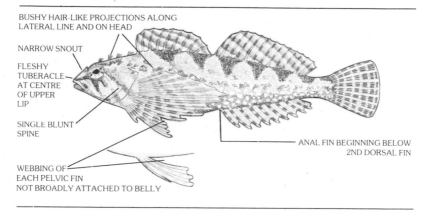

BUSHY HAIR-LIKE PROJECTIONS ALONG
LATERAL LINE AND ON HEAD

NARROW SNOUT

FLESHY
TUBERACLE
AT CENTRE
OF UPPER
LIP

SINGLE BLUNT
SPINE

ANAL FIN BEGINNING BELOW
2ND DORSAL FIN

WEBBING OF
EACH PELVIC FIN
NOT BROADLY ATTACHED TO BELLY

SPECIES: _Clinocottus embryum_—from the Greek _clinus_, _cottus_, _en_ and _bryon_, meaning "blenny," "sculpin," "in," and "sea moss."

ALTERNATE NAMES: mossy sculpin, bullhead*.

MAXIMUM RECORDED SIZE: 7 cm (2.8 inches).

DISTRIBUTION: northern Baja California, Mexico, to the Aleutian Islands, and the Bering Sea coast of Alaska.

The small size of the calico sculpin and its shallow, surf-swept environment effectively protect it from most anglers.

The constant surging and crashing of waves and associated poor visibility in the shallow or intertidal habitat of the tiny calico sculpin discourage most divers from searching for this species. "Calico" beautifully describes the intricate colour pattern of this pretty sculpin, a creature difficult to see when it huddles among submerged vegetation. A snorkeler might have some luck searching carefully in a large rocky-bottom tidepool.

Commercial fishermen seldom seek any quarry within the calico sculpin's intertidal habitat.

Commonly found in rocky, weed-choked tidepools of nearly any size, the secretive calico sculpin lives only along shorelines directly swept by the Pacific and usually lurks amid the branches of pink corralline algae. Wait patiently for a hungry specimen to dart after a tiny shrimp or other creatures and betray itself. Difficult to distinguish from several other intertidal sculpins, the well-camouflaged calico makes an interesting inmate for an all-glass marine aquarium; in captivity, in chilled seawater with an adequate supply of live food, this sculpin may exhibit some fascinating feeding, mating, or territorial behaviours. Take a few home in a sturdy plastic bag resting on some ice.

Too tiny.

*—incorrect

139

106 Mosshead Sculpin

DENSE BUSHY HAIR-LIKE PROJECTIONS
ON FORWARD PART OF LATERAL LINE

2ND DORSAL FIN

BLUNT HEAD

SINGLE BLUNT
SKIN-COVERED SPINE

ANAL FIN BEGINS
BELOW 2ND DORSAL FIN

SPECIES: *Clinocottus globiceps*—from the Greek *clinus* and *cottus*, meaning "blenny" and "sculpin;" and the Latin *globus* and *ceps*, connoting "globe" and "head."

ALTERNATE NAMES: globe-headed sculpin, round-headed sculpin, bullhead *.

MAXIMUM RECORDED SIZE: 19 cm (7.5 inches).

DISTRIBUTION: Graviota, southern California, to Kodiak Island, Gulf of Alaska.

Only by an unlikely accident would a bait-fisherman catch the primarily intertidal mosshead sculpin along rocky, weed-covered and surf-swept shallows. If such an "accident" occurs, use the catch as bait for larger bottomfish.

An intertidal creature nestling in the crevices of large rock formations, the boldly marked mosshead sculpin sits very still much of the time and therefore may be difficult for divers to observe. Aquanauts snorkeling in large tidepools sheltered from direct crashing surf have the best opportunity to study this creature; bits of smashed barnacle or mussel may lure it out into the open.

Commercial fishermen do not harvest the shallow habitat of the shelter-seeking mosshead sculpin.

Look for this distinctive sculpin in the middle to lower intertidal zone, but only along shores exposed directly to Pacific waves. If for some reason it is displaced from its particular tidepool, this fish can eventually find its way back. Once, while collecting for the Vancouver Public Aquarium early one still morning near Tofino on Vancouver Island, we collectors were surprised by a splashing sound from several nearby tidepools, and after quietly approaching the nearest pool were amazed to find several mosshead sculpins perched on the rocks and completely out of the water! Once disturbed, though, these amphibious creatures retreated quickly, splashing back into the water.

Theoretically, a large mosshead sculpin possesses enough flesh for a taste.

*—incorrect

140

107 Rosylip Sculpin

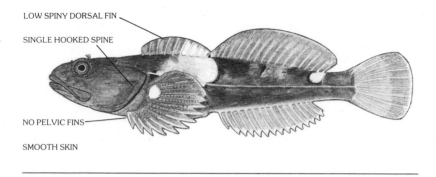

LOW SPINY DORSAL FIN

SINGLE HOOKED SPINE

NO PELVIC FINS

SMOOTH SKIN

SPECIES: *Ascelichthys rhodorus*—from the Greek *a*, *scelos*, *ichthys*, *rodor* and *orus*, meaning "without," "legs," "fish," "rose," and "margin."

ALTERNATE NAMES: rosy-lipped sculpin, bullhead*.

MAXIMUM RECORDED SIZE: 15 cm (6 inches).

DISTRIBUTION: San Mateo, central California, to Sitka, southeastern Alaska.

Although the slow-moving rosylip sculpin thrives in shallow shoreline habitats where anglers frequently bait-fish, it seldom seizes their offerings. If an angler could gather enough specimens, this bottom dweller might tempt a large greenling, rockfish or flounder.

The smooth-skinned rosylip sculpin inhabits gravel beaches, rocky shorelines or eelgrass-choked bays at depths of less than 10 m (33 feet). In such generally murky locales, its dark coloration and its habit of huddling motionless amid surrounding shelter combine to conceal it from even the most observant diver. Occasionally a diver may have a quick glimpse of a hungry specimen as it bolts from cover to seize a tiny shrimp or as a frightened one flees from a patrolling aquanaut.

Few commercial harvesters venture into the shallow domain of the unsaleable rosylip sculpin and so this fish probably never sees the deck of a fish boat.

Rocky tidepools, even very tiny ones or those located high in the intertidal zone, may contain the dark rosylip sculpin—though a beachcomber must look closely to see it because it usually remains very still, nestling upon the bottom contours. Young specimens, less than 2.5 cm, are common, particularly along gravel beaches or among adjacent weeds, in late spring. The distinctive rosylip is one of the easiest small shallow water sculpins to identify because it is the only one without pelvic fins!

Perhaps some desperate soul has actually eaten the small rosylip sculpin—but who?

*—incorrect

108 Prickly Sculpin

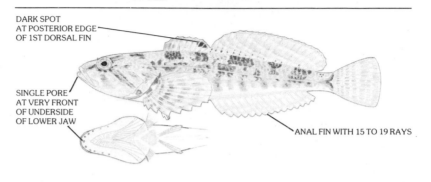

DARK SPOT
AT POSTERIOR EDGE
OF 1ST DORSAL FIN

SINGLE PORE
AT VERY FRONT
OF UNDERSIDE
OF LOWER JAW

ANAL FIN WITH 15 TO 19 RAYS

SPECIES: *Cottus asper* —from the Greek *cottus* and *asper*, meaning "sculpin" and "rough."

ALTERNATE NAMES: prickly bullhead*, bullhead*.

MAXIMUM RECORDED SIZE: 30 cm (12 inches).

DISTRIBUTION: Ventura River, central California, to Seward, on the Kenai Peninsula, Alaska.

Anglers who still-fish with bait along the sand bars of coastal river estuaries often catch the sluggish prickly sculpin, considered a time-wasting nuisance because it usually engulfs worms, roe, and other baits set for more desirable salmon or trout. That aquatic birds, squawfish, and even trout prey upon it indicates some bait potential. Spawning occurs in late winter and spring when mature adults move down stream, often into brackish water where each female lays as many as 10,000 eggs under a rock or even in a tin can. There are several races of prickly sculpin, however, and the one without prickles on its skin lives along the coast and occasionally invades salt water.

Only an occasional intruder into the marine environment, the pugnacious prickly sculpin usually stays reasonably close to stream mouths, living at depths less than 10 m (33 feet). Search for it in and around river estuaries, the waters of which are often murky.

The unsaleable prickly sculpin slips easily through the large meshes of salmon gillnets, the commercial gear most often fished in river mouths.

Look along the banks of stream or river estuaries to find the grayish-brown prickly sculpin lurking upon gravelly and sandy bottoms. This common species moves periodically to search for prey such as migrating salmon fry which pass through its territory while it avoids its own numerous predators.

Soften some cheese; add horseradish, onion, lemon and mashed, de-boned prickly sculpin. Shape mixture on a plate and spread it with parsley. Use as a pate.

109 Manacled Sculpin

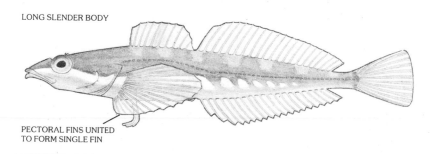

LONG SLENDER BODY

PECTORAL FINS UNITED
TO FORM SINGLE FIN

SPECIES: *Synchirus gilli*—from the Greek *syn* and *cheir*, meaning "together" and "hand;" and honouring early American ichthyologist Theodore Gill.

ALTERNATE NAMES: none.

MAXIMUM RECORDED SIZE: 7 cm (2.8 inches).

DISTRIBUTION: San Miguel Island, southern California, to Sitka, southeastern Alaska.

Even as bait, the tiny manacled sculpin is useless to the angler.

In shallows, search through the various plants growing in kelp forests or attached to floats to see the greenish, golden-brown or mauve manacled sculpin. Between short bursts of activity this little predator of tiny shrimps and their kin rests upon fronds of kelp or other weed and a diver must look very closely to see it. Taking a well-lighted, correctly centred and clearly focused photograph of the wary manacled sculpin offers a great challenge to the underwater photographer using a macro lens.

Small enough to pass through the finest commercial net meshes set in its domain, the inconspicuous manacled sculpin never enters commercial catches.

Among the sea lettuce, kelp, and other algae that hang from current-swept floats, piers, or pilings, the slender manacled sculpin swims within easy viewing distance. Seldom is it trapped in tidepools. The male manacled sculpin has a very prominent silvery lateral stripe on each side, small eyes, large pelvic fins, a long jaw with a definite knob on its chin and a genital papilla. During the winter and spring breeding season, watch the aggressive male as he finds a prospective mate and grips her with his specialized pelvic fins and lower jaw before inserting his genital papilla to fertilize her eggs internally. So tight is their embrace that a naturalist may lift the stimulated pair clear out of the water without the tenacious male dropping his consort. Later this gravid female will deposit small pinkish adhesive eggs upon some submerged solid object, particularly on the inside of kelp holdfasts (root-like structures).

Too tiny for the kitchen.

110 Longfin Sculpin

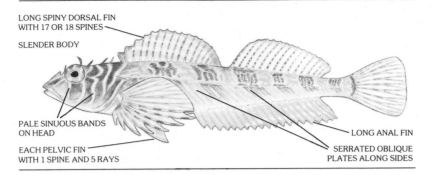

LONG SPINY DORSAL FIN
WITH 17 OR 18 SPINES

SLENDER BODY

PALE SINUOUS BANDS
ON HEAD

EACH PELVIC FIN
WITH 1 SPINE AND 5 RAYS

LONG ANAL FIN

SERRATED OBLIQUE
PLATES ALONG SIDES

SPECIES: *Jordania zonope*—honouring David Starr Jordan, regarded as the father of American ichthyology; and from the Greek *zone* and *ope*, meaning "zone" and "window."

ALTERNATE NAMES: bandeye sculpin.

MAXIMUM RECORDED SIZE: 15 cm (6 inches).

DISTRIBUTION: Point Lobos, central California, to Port Walker, Baronoff Island, southeastern Alaska.

Though it might nibble at the bait, the tiny longfin sculpin can never hook its minute mouth on conventional angling tackle. A potential bait source for bottomfish; rockfishes and greenlings prey on it.

Scientists seldom observed the longfin sculpin before exploring with SCUBA became popular. The shallow rocky reef habitat of this brighly-hued creature makes it a common find for divers cruising depths less than 20 m (66 feet). Look among the caves and crevices for this sculpin, which seems constantly moving in fits and starts, but, when it is momentarily stationary, rests propped upon its fanlike pectoral fins. On other occasions it might linger on vertical cliffs or even "hang" upside down and horizontally along a cavern roof. Following an elaborate series of courtship manoeuvres, a gravid female longfin sculpin lays several small clusters of amber-coloured eggs which the darkened male doggedly guards until they hatch.

Only very rarely does a commercial prawner pull up a trap containing a longfin sculpin—an unwanted incidental. Depths as great as 38 m (125 feet) provide habitat for this species.

At very extreme low tides, explore rocky-bottom tidepools for a glimpse of the longfin sculpin. Persevere, though, for such sightings are uncommon.

Difficulty of capture and size of the species eliminate the longfin sculpin as a menu item.

144

111 Spinynose Sculpin

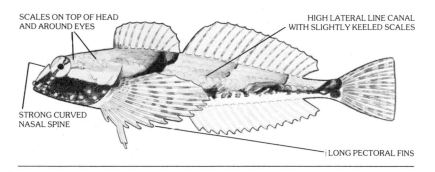

SCALES ON TOP OF HEAD
AND AROUND EYES

HIGH LATERAL LINE CANAL
WITH SLIGHTLY KEELED SCALES

STRONG CURVED
NASAL SPINE

LONG PECTORAL FINS

SPECIES: *Asemichthys taylori*—from the Greek *a*, *sem*, and *ichthys*, meaning "negative," "sign," and "fish;" as well as honouring George W. Taylor, the first director of the Pacific Biological Station, Nanaimo, British Columbia.

ALTERNATE NAMES: Taylor's sculpin.

MAXIMUM RECORDED SIZE: 5.7 cm (2.3 inches).

DISTRIBUTION: San Juan Island, northern Washington, to Tasu Harbour, Queen Charlotte Islands, northern British Columbia.

The tiny spinynose sculpin is too small to be a practical bait for large bottomfish.

Easily overlooked by most divers, the slender elongate spinynose sculpin usually rests on or actually buries itself partially in the shell hash which gathers in level terraces of reefs and around the bases of rocky outcroppings below 10 m (33 feet). Difficult to see unless it moves, the sometimes reddish-pink specimens lurk among the red algae attached to rocky shelves adjacent to favoured shell debris. Until the advent of SCUBA, scientists seldom saw the elusive spinynose sculpin, so much of its natural history and geographic distribution still remains unclear. Once able to identify this inconspicuous little fish, log information about its habits, behaviour, and distribution: such first-hand data adds a great deal to a dive log.

Obviously the active spinynose sculpin is too small for the market and it is rarely caught. Even though it lives on rocky bottoms at depths to 50 m (165 feet), where prawn traps might be set, this wary fish slips easily through the tiny meshes.

Because much remains to be discovered about the seldom-seen spinynose sculpin, some surface-bound naturalist may be the first to observe and record it intertidally or from a wharf.

Too small for eating.

112 Northern Sculpin

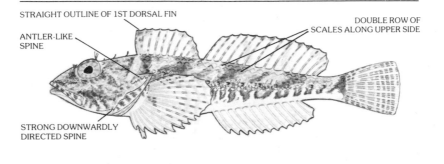

STRAIGHT OUTLINE OF 1ST DORSAL FIN

ANTLER-LIKE SPINE

DOUBLE ROW OF SCALES ALONG UPPER SIDE

STRONG DOWNWARDLY DIRECTED SPINE

SPECIES: *Icelinus borealis*—from the Greek *Hicelos* meaning "God of sleep;" and the Latin *borealis*, "northern."

ALTERNATE NAMES: comb sculpin, bullhead*.

MAXIMUM RECORDED SIZE: 10.2 cm (4 inches).

DISTRIBUTION: Puget Sound, northern Washington, to the Bering Sea coast of Alaska, and Attu Island, in the Aleutian chain.

Even if the small northern sculpin were to nibble at baits intended for bigger, desirable quarry, this fish is unlikely to hook itself upon any large commonly used tackle.

The drab northern sculpin frequently lives on the bottom at depths below 10 m (33 feet) and prefers to sit upon large, smooth rocky outcroppings or barren reefs that are covered by silt but little else. Its tiny brownish-gray form usually remains unnoticed unless it stirs up silt as it moves. When the well camouflaged northern sculpin chases shrimp-like prey, an observant diver can most easily view it during its frequent rest stops in the beam of an underwater flashlight. During night dives, too, the stalking aquanaut may also encounter this common but little-studied creature.

Enticed by captive prawns and smaller shrimps, hungry northern sculpins often find their way into prawn traps. Although this common sculpin is too tiny to sell, shrimp trawlers sometimes scoop it up and then drag it to the boat from level silty or smooth rocky bottoms at depths between 10 and 220 m (33 and 726 feet).

Occasionally the primarily bottom-dwelling northern sculpin, attracted by dock or jetty lights, may swim to the surface where it might be observed. Intriguingly, plankton researchers sometimes collect the tiny transparent larvae of this fish hundreds of miles from shore.

Too small for eating.

*—incorrect

146

113 Dusky Sculpin

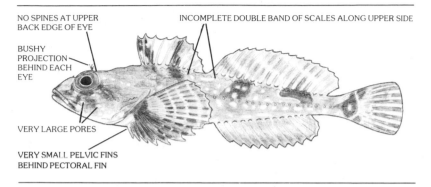

NO SPINES AT UPPER
BACK EDGE OF EYE

INCOMPLETE DOUBLE BAND OF SCALES ALONG UPPER SIDE

BUSHY
PROJECTION
BEHIND EACH
EYE

VERY LARGE PORES

VERY SMALL PELVIC FINS
BEHIND PECTORAL FIN

SPECIES: Icelinus burchami—from the Greek *Hicelos* meaning "God of sleep;" and honouring James S. Burcham, an early member of the United States Bureau of Fisheries.

ALTERNATE NAMES: bullhead *.

MAXIMUM RECORDED SIZE: 13 cm (5 inches).

DISTRIBUTION: San Nicholas Island and La Jolla, southern California, to Loring, southeastern Alaska.

Few bait-fishermen lower their tackle to the dark, silty-bottomed habitat of the small dusky sculpin. Although this wary fish might prove somewhat useful as bait, anglers have no access to it and so they cannot even try it.

Current records define the grayish dusky sculpin's minimum depth to be 61 m (200 feet), relegating it to very doubtful status on any diver's list of viewable fishes. However, as SCUBA aficionados descend into virgin dive locales, some intrepid explorer may yet find a population of dusky sculpins living upon shallower silty or sandy bottoms. Pilots and passengers of submersibles are probably the underwater investigators most likely to encounter this tiny, rarely-seen species.

Sometimes shrimp trawlers who tow their gear over level bottoms between depths of 70 and 570 m (330 and 1,880 feet) inadvertently catch the unmarketable dusky sculpin. In search of an easy meal, this sluggish fish may also enter baited prawn traps set at these depths. While too small to be of direct economic value, the inconspicuously-coloured dusky sculpin is undoubtedly a dietary item for some commercially valuable species, including the various flounders and codfishes, as well as the Pacific hake.

The bottom-dwelling dusky sculpin flourishes in deep water, inaccessible to the pierside naturalist, beach stroller or boating enthusiast.

Too small for serious eating.

* —incorrect

147

114 Spotfin Sculpin

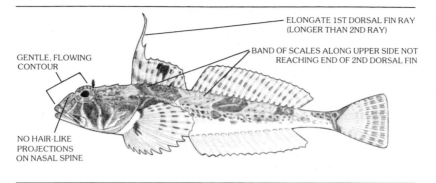

ELONGATE 1ST DORSAL FIN RAY
(LONGER THAN 2ND RAY)

BAND OF SCALES ALONG UPPER SIDE NOT
REACHING END OF 2ND DORSAL FIN

GENTLE, FLOWING
CONTOUR

NO HAIR-LIKE
PROJECTIONS
ON NASAL SPINE

SPECIES: *Icelinus tenuis*—from the Greek *Hicelos* meaning "God of sleep;" and the Latin *tenuis*, "slender."

ALTERNATE NAMES: lesser filamented sculpin, bullhead*.

MAXIMUM RECORDED SIZE: 14 cm (5.5 inches).

DISTRIBUTION: San Benito Island, central Baja California, Mexico, to Tasu Sound, Queen Charlotte Islands, northern British Columbia, and perhaps Alaska.

Few anglers encounter the small spotfin sculpin, a tiny-mouthed fish that seldom seizes baits of any type. If specimens could be acquired, though, they might prove valuable as bait for large-mouthed flounders or other sandy-bottom dwellers.

Often while searching for prawns over sandy bottoms at depths below 15 m (50 feet) night divers spy the slender spotfin sculpin illuminated in the beams of their underwater flashlights. Either huddled belly down on the sand or propped up on its pectoral fins, this nocturnal fish with its two incised first dorsal fin rays, waits motionless for small shrimps or other prey. Its pale grayish brown colour may conceal it not only from various predators but also from the unobservant diver. Occasionally, keen-eyed aquanauts find this drab sculpin on a sortie during a dark overcast day.

Shrimp trawlers who drag their fine-mesh nets over flat, soft bottoms at depths to 370 m (1,220 feet) sometimes capture the commercially valueless spotfin sculpin. Attracted by a possible meal, this opportunistic fish occasionally enters prawn traps set upon its level-bottomed habitat. Surprisingly, experimental mid-water trawls made by Fisheries and Oceans Canada have captured the primarily bottom-dwelling spotfin sculpin well above the sea floor, where the fish is vulnerable to completely different predators.

The young spotfin sculpin could possibly be attracted to artificially lighted docks or jetties where nocturnal naturalists could see it.

Too small for dining.

*—incorrect

115 Threadfin Sculpin

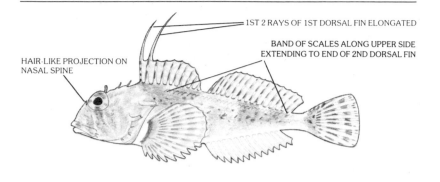

1ST 2 RAYS OF 1ST DORSAL FIN ELONGATED

BAND OF SCALES ALONG UPPER SIDE
EXTENDING TO END OF 2ND DORSAL FIN

HAIR-LIKE PROJECTION ON
NASAL SPINE

SPECIES: *Icelinus filamentosus*—from the Greek *Hicelos* meaning "God of sleep;" and the Latin *filamentosus*, "thread."

ALTERNATE NAMES: filamented sculpin, long-rayed sculpin, bullhead*.

MAXIMUM RECORDED SIZE: 27 cm (10.7 inches).

DISTRIBUTION: Cortez Bank, near Point Loma, southern California, to Vancouver Island, British Columbia, and perhaps Alaska.

While angling to depths below 37 m (122 feet) for more desirable quarry, bait-fishers may perhaps take the unwanted threadfin sculpin. They should not discard it but instead, impale it on a large hook, lower it back down alive and continue fishing.

As yet the bulbous-headed threadfin sculpin has only been found in water as shallow as 37 m (122 feet), just below depths explored by most sport divers. However, it is conceivable that this deep water creature may invade shallower depths and a diver may very well discover one on a muddy or sandy bottom. While the threadfin sculpin awaits its next meal of shrimps or shrimp-like animals, its distinctive silhouette would be very easily noticed as it rests upon the sea floor.

The unsaleable threadfin sculpin is only an incidental curiosity or nuisance to the commercial shrimp trawler who occasionally drags it from even bottoms at depths to 373 m (1,224 feet). After finding this opportunistic creature trapped in their gear, resourceful prawn trappers could kill it, tie it to the bait container, and then return the freshly-baited trap to the sea floor.

Surface-bound naturalists hardly ever see the deep water threadfin sculpin unless it is in a public aquarium.

Slice threadfin sculpin fillets into small pieces, simmer with carrots, then add water, potato, green beans, stewed tomatoes, milk, onion, salt and thyme before simmering further for a delicious soup.

*—incorrect

116 Ribbed Sculpin

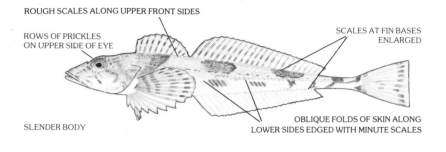

ROUGH SCALES ALONG UPPER FRONT SIDES

ROWS OF PRICKLES
ON UPPER SIDE OF EYE

SCALES AT FIN BASES
ENLARGED

SLENDER BODY

OBLIQUE FOLDS OF SKIN ALONG
LOWER SIDES EDGED WITH MINUTE SCALES

SPECIES: *Triglops pingeli*—from *trigla* the Greek word for the somewhat similar-looking gurnard, and *ops*, meaning "like;" and honouring an unknown individual.

ALTERNATE NAMES: none.

MAXIMUM RECORDED SIZE: 20 cm (8 inches).

DISTRIBUTION: Washington to the Bering Sea, then west through the Aleutian chain to Hokkaido, northern Japan as well as east through the Arctic to Hudson's Bay, Cape Cod and Denmark.

A small, slender, tapered species, the unwanted ribbed sculpin bothers very few anglers.

Unless exploring flat, uninteresting mud and gravel bottoms, the casual diver will not likely encounter or recognize the elusive ribbed sculpin. To identify the species two divers captured several specimens, but only after considerable effort; the active fish cruised along, usually just beyond the collector's reach, and seldom stopped to rest. Eventually, by careful team stalking and slow deliberate work with a hand net, the crew obtained several specimens. They were transported alive to the West Vancouver Laboratory by sealing the healthy captives in a plastic bag containing a small amount of seawater and some pure oxygen, and bringing them back to the laboratory on ice in a picnic cooler. A diver can obtain oxygen from a specially purchased cylinder through plastic tubing connected to an oxygen distribution regulator.

While after shrimps, trawlers sometimes drag unmarketable ribbed sculpins from soft level substrates at depths between 12 and 72 m (40 and 304 feet).

Young colourful ribbed sculpin often live in and around the eelgrass beds at depths as shallow as 3 m (10 feet).

Too slender and with too little flesh for serious eating.

117 Roughspine Sculpin

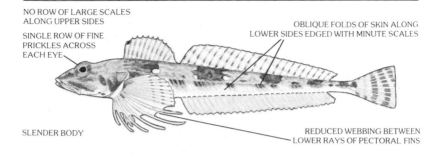

NO ROW OF LARGE SCALES
ALONG UPPER SIDES

SINGLE ROW OF FINE
PRICKLES ACROSS
EACH EYE

OBLIQUE FOLDS OF SKIN ALONG
LOWER SIDES EDGED WITH MINUTE SCALES

SLENDER BODY

REDUCED WEBBING BETWEEN
LOWER RAYS OF PECTORAL FINS

SPECIES: ***Triglops macellus***—from *trigla* the Greek word for the somewhat similar-looking gurnard, and *ops*, meaning "like;" as well as the Latin *macer* and *ella*, "thin" and "tiny."

ALTERNATE NAMES: rough-spine sculpin.

MAXIMUM RECORDED SIZE: 20 cm (8 inches).

DISTRIBUTION: Washington to the Bering Sea coast of Alaska.

A small fish with a tiny mouth, the seldom-noticed roughspine sculpin is unlikely to encounter the angler's hook.

Watch for the nocturnally active roughspine sculpin, a slender and tapered fish which forages actively for small invertebrate prey and often rests on the bottom, perched upon its pectoral and tail fins, looking like a tiny single-engine airplane awaiting take-off. The underwater photographer may then take a good picture of the creature highlighted by a lack of background instead of it being obscured by the nondescript bottom. On other occasions the brownish-gray roughspine sculpin may fold its pectoral fins, huddle belly-down on the sand and remain quite inconspicuous. A diver moving cautiously can easily approach any specimen in that position without making it take off into the dark. Look for roughspine sculpin, particularly upon sandy bottoms, well clear of any rocky outcroppings, at depths below 15 m (50 feet).

Although sometimes scooped up in the fine mesh of a shrimp trawler's gear, the brownish roughspine sculpin is an unprofitable nuisance, but undoubtedly has some incidental value as prey for other large valuable groundfish that share its habitat and live down to 92 m (303 feet).

Beachcombers and dockside strollers are unlikely to observe the wary roughspine sculpin.

Too slender.

118 Slim Sculpin

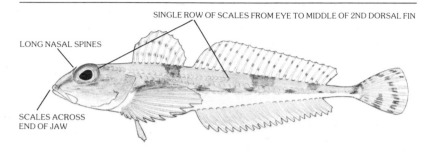

SINGLE ROW OF SCALES FROM EYE TO MIDDLE OF 2ND DORSAL FIN

LONG NASAL SPINES

SCALES ACROSS
END OF JAW

SPECIES: *Radulinus asprellus*—from the Latin *radula, asper* and *ella,* meaning "scraper," "rough," and "tiny."

ALTERNATE NAMES: darter sculpin *.

MAXIMUM RECORDED SIZE: 15 cm (6 inches).

DISTRIBUTION: northern Baja California, Mexico to Amchitka Island in the Aleutian chain, Alaska.

The seldom seen slim sculpin is too small to be caught with nearly any hook size, but potentially could serve as bait for large flounders or codfishes.

While patrolling deep muddy bottoms, particularly at night, an observant diver might find the light grayish slim sculpin huddling there. An underwater photographer who finds one should approach it from down-current, so that particulate matter does not drift over and obscure the specimen, thus spoiling a noteworthy photo.

Although of no direct commercial value, the slim sculpin is a common incidental catch for shrimp trawlers who drag their nets over soft, level substrates. This inconspicuous sculpin, which undoubtedly falls prey to large, commercially important groundfish, illustrates an interesting geographic-depth relationship also exhibited by other marine fishes of North America's western shore: within the boundaries of the known range of the slim sculpin, the farther south collectors have found it, the more likely it will have come from deep water. In California, for example, scientists have captured it at depths as great as 284 m (930 feet), while specimens in British Columbia have been gathered at a maximum of 180 m (594 feet). Consequently, scientists believe that the slender slim sculpin has a preferential temperature range, and these temperatures occur deeper in more southerly latitudes.

As yet no documentation from shallow or intertidal habitats exists for the wary slim sculpin but perhaps some future beachcomber or dockside observer will be able to supply that data.

Too slender for serious eating.

* —incorrect

119 Darter Sculpin

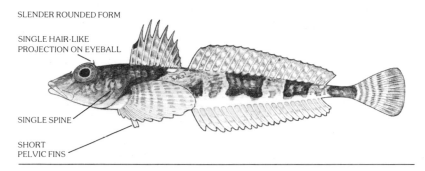

SLENDER ROUNDED FORM

SINGLE HAIR-LIKE
PROJECTION ON EYEBALL

SINGLE SPINE

SHORT
PELVIC FINS

SPECIES: *Radulinus boleoides*—from the Latin *radula*, meaning "scraper;" and the Greek *boleo* and *oid*, "dart" and "like."

ALTERNATE NAMES: none.

MAXIMUM RECORDED SIZE: 14 cm (5.5 inches).

DISTRIBUTION: Santa Catalina Island, southern California, to Langara Island in the Queen Charlotte Islands, northern British Columbia.

The small seldom-encountered darter sculpin lives at depths well below those usually fished by bait offering anglers who might catch it on tiny hooks. If available, this slender sculpin might make a useful bait for bottomfishermen.

During a memorable September 1981 night dive in Howe Sound, a party of divers at about 21 m (70 feet) happened upon what was obviously an unusual fish resting upon the steeply sloping, muddy and debris littered bottom. Bernie Hanby, an amateur underwater photographer, moved in and fired off about ten frames; we had to illuminate the subject while taking care not to stir up the loose bottom. Then we had to be slow and deliberate. I lowered my open hand onto the co-operative creature, grasped quickly, picked it up, and placed it in an empty bleach bottle, the only means we had of transporting it back to the laboratory. After identifying the creature as a darter sculpin, we sketched it and re-photographed it: the fifth such catch recorded in British Columbia, and the first by divers.

Shrimp trawlers dragging their fine-meshed nets over soft substrates between 21 and 146 m (70 and 480 feet) might inadvertently capture the unmarketable darter sculpin.

The grayish darter sculpin apparently lives at depths inaccessible to surface-bound naturalists.

All that have been caught in British Columbia would hardly make a panful.

120 Roughback Sculpin

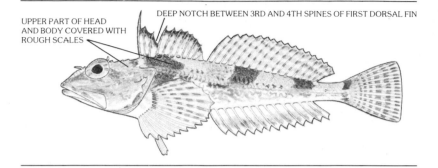

UPPER PART OF HEAD AND BODY COVERED WITH ROUGH SCALES

DEEP NOTCH BETWEEN 3RD AND 4TH SPINES OF FIRST DORSAL FIN

SPECIES: *Chitonotus pugetensis*—from the Greek *chiton* and *notus*, meaning "tunic" and "back;" and emphasizing Puget Sound.

ALTERNATE NAMES: bullhead *.

MAXIMUM RECORDED SIZE: 23 cm (9 inches).

DISTRIBUTION: Bahia Santa Maria, southern Baja California, Mexico, to Trail Island, Wark Channel, northern British Columbia.

Baitfishermen do not seek the slender roughback sculpin but at night it sometimes seizes small offerings lowered from piers or floats.

A very common sight for prawn-seeking night divers, the grayish roughback sculpin flourishes upon sandy bottoms at depths below 9 m (30 feet). By wriggling its body to and fro and scooping with its fan-like pectoral fins, this distinctively silhouetted creature often buries itself in the mud or sand, leaving only its back and large eyes exposed. Carefully dig one out and watch the unstartled specimen swim a short distance away and then bury itself again. As its long breeding season approaches, a mature male often darkens, the irregular red bands on his sides intensify in colour and he may even develop a bright red, orange or white flash on top of each eye. After being courted by such a "dashing" suitor, the gravid female deposits clusters of fertilized, bright salmon-coloured eggs on the sea floor.

A small but unprofitable catch for the shrimp trawler or prawn trapper, the common roughback sculpin nonetheless shows up in hauls. Depths to 140 m (462 feet) provide habitat for this inconspicuous fish which preys heavily upon small shrimps and shrimp-like creatures.

Although generally lurking in deep water, the nocturnally active roughback sculpin may approach lights which shine into the water from a float or pier.

There is only a minute amount of edible flesh on the largest roughback sculpin.

*—incorrect

121 Spinyhead Sculpin

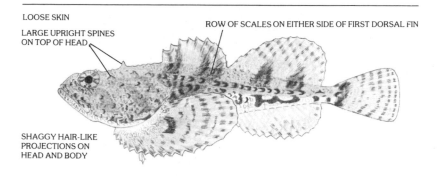

LOOSE SKIN

LARGE UPRIGHT SPINES
ON TOP OF HEAD

ROW OF SCALES ON EITHER SIDE OF FIRST DORSAL FIN

SHAGGY HAIR-LIKE
PROJECTIONS ON
HEAD AND BODY

SPECIES: *Dasycottus setiger* —from the Greek *dasy* and *cottus*, meaning "warty" and "sculpin;" as well as the Latin *setos* and *gero*, connoting "bristle" and "to bear."

ALTERNATE NAMES: wooly sculpin*, bullhead*.

MAXIMUM RECORDED SIZE: 23 cm (9 inches).

DISTRIBUTION: Puget Sound, northern Washington, to the Aleutian chain, the Bering Sea and Honshu Island, central Japan.

The lethargic spinyhead sculpin inhabits deep water and silty bottoms where baitfishermen do not usually fish. If a specimen were captured and if it were lively enough, a fisherman could perhaps re-hook it and lower it back down as bait.

The bright beam of a night diver's flashlight sometimes illuminates the spinyhead sculpin, with its warty head, as it nestles upon sandy or silty bottoms at depths below 15 m (50 feet). Its dull pink-to-gray colour provides camouflage while it hides in drab muddy surroundings, particularly among scattered rock. Once located, the inconspicuous spinyhead sculpin often remains very still and may even submit to gentle handling for a posed photograph. This species has bristles or cirri on its head and the larger the specimen the more cirri and the more bristly the appearance.

Shrimp trawlers and, to a lesser extent, prawn trappers who fish their gear on sandy or silty substrates at depths to 140 m (462 feet), often catch the unprofitable spinyhead sculpin as an incidental. Those shrimps and prawns so eagerly sought are also prey gulped by this lurking predator.

Deep water habitats that shelter the drab spinyhead sculpin also isolate it from surface-bound naturalists.

Because of the modest quantity and flabby nature of its flesh, even a large spinyhead sculpin interests few seafood fanciers.

*—incorrect

155

122 Blackfin Sculpin

LOOSE SKIN

NO SPINES ON TOP
OF HEAD

LIGHT PECTORAL FINS,
DARKLY MARGINED

SPECIES: *Malacocottus kincaidi*—from the Greek *malaco* and *cottus*, meaning "soft" and "sculpin;" and honouring Professor Trevor Kincaid of the University of Washington.

ALTERNATE NAMES: bullhead*.

MAXIMUM RECORDED SIZE: 20 cm (8 inches).

DISTRIBUTION: Washington to Bird Island off the Alaskan Peninsula, southeastern Alaska, and recorded in Japan.

A small species, generally inhabiting deep water, the globular blackfin sculpin seldom takes the angler's offerings. Less often is it actually landed. Perhaps a useful bait.

Sometimes while night diving over silty bottoms where boulders lie scattered about, an observant aquanaut may encounter a stationary blackfin sculpin. Expect to find it at depths below 18 m (60 feet) but search patiently because the globular head and brownish coloration of this creature make it fairly difficult to spot. Certain blackfin sculpins, as well as several other deep water, bottom-dwelling fishes of the Pacific Northwest, have beautiful bright blue pectoral fins which not only assist the underwater explorer in finding them, but also attract attention from avid underwater photographers.

The unmarketable, stubby blackfin sculpin is quite often swept into the fine-meshed nets of shrimp trawlers dragging along soft, level bottoms at depths to 275 m (902 feet). Disgruntled fishermen must laboriously sort the profitable shrimp from the remaining unwanted fishes or invertebrates, and usually pitch the soft blackfin sculpin overboard. Surprisingly, scientific expeditions using mid-water trawls have captured this slow-moving creature at between 90 and 275 m (300 and 908 feet), well above the sea floor, 2,000 m (6,600 feet) below.

Don't expect to see the deep water blackfin sculpin from any above-surface vantage point.

The flabby body of the smallish blackfin sculpin is not appealing.

* —incorrect

123 Tadpole Sculpin

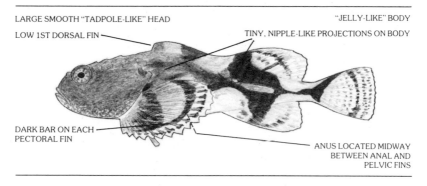

LARGE SMOOTH "TADPOLE-LIKE" HEAD

LOW 1ST DORSAL FIN

"JELLY-LIKE" BODY

TINY, NIPPLE-LIKE PROJECTIONS ON BODY

DARK BAR ON EACH PECTORAL FIN

ANUS LOCATED MIDWAY BETWEEN ANAL AND PELVIC FINS

SPECIES: *Psychrolutes paradoxus*—from the Greek *psychrolutes* and *paradoxus*, meaning "one who bathes in cold water" and "unaccountable."

ALTERNATE NAMES: none.

MAXIMUM RECORDED SIZE: 6.4 cm (2.5 inches).

DISTRIBUTION: Puget Sound, northern Washington, to the Bering Sea and through to the Sea of Japan.

The almost jelly-like body of the tiny tadpole sculpin makes it a doubtful bait, perhaps one avoided by larger fish.

Resting upon silty bottoms at depths to 220 m (720 feet), the brownish tadpole sculpin, with its black-edged and often orange-centred pectoral fins is abundant in certain shallow areas, particularly in summer. Frequently this slow swimmer huddles among the eelgrass or rests upon the surrounding mud but, in spite of its small size, most divers may easily see it. Minute tadpole sculpin larvae live briefly at the surface during summer. These young drift with the current and feed on free-swimming barnacle larvae.

Only the very fine mesh trawls of the commercial shrimper retain the largest tadpole sculpins as unwanted incidentals. Experimental midwater trawls have actually gathered the limp but living carcasses of this weak swimmer well above the sea floor.

Tadpole sculpins commonly swarm into muddy, shallow bays where a keen-eyed naturalist may see them from piers or jetties in spring and summer. At night, though, an observer may see them more easily because they readily swim off the bottom toward beams of light. Dipnet a specimen, lift it from the water, and notice that its flaccid body appears to be an inert glob of jelly.

The texture of the tiny tadpole sculpin is most unappetizing.

124 Soft Sculpin

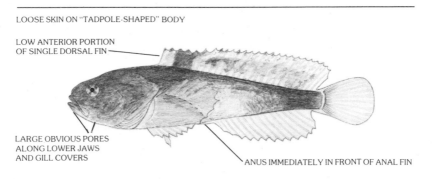

LOOSE SKIN ON "TADPOLE-SHAPED" BODY

LOW ANTERIOR PORTION
OF SINGLE DORSAL FIN

LARGE OBVIOUS PORES
ALONG LOWER JAWS
AND GILL COVERS

ANUS IMMEDIATELY IN FRONT OF ANAL FIN

SPECIES: *Gilbertidia sigalutes*—honouring Charles Gilbert, noted early American ichthyologist; as well as from the Latin *sig* and *lutes*, meaning "quiet" and "bather."

ALTERNATE NAMES: none.

MAXIMUM RECORDED SIZE: 8.3 cm (3.3 inches).

DISTRIBUTION: Washington to the Aleutian Island chain, Alaska.

The globular soft sculpin is too small to take almost any hook and its jelly-like body seems to discourage some large fish from preying upon it.

In certain areas, a night diver might notice the slow-swimming soft sculpin in shallow waters around floats. Look for its grayish form over silty bottoms adjacent to eelgrass beds where the sculpin often rises off the bottom and swims toward artificial light. Some reports indicate this active fish might also seek shelter inside sponges.

Initially lost among the writhing mass of shrimps and other creatures dragged aboard trawlers from depths to 225 m (743 feet), the smooth soft sculpin is obviously too small for market.

Dr. Charles Moffett, a psychiatrist from Friday Harbour, Washington, enthusiastically greets the springtime arrival of the soft sculpin. On still, windless evenings he watches for this abundant, slow-swimming fish to approach his "night light" which hangs from a neighbour's float. Occasionally forming distinct schools, a behaviour very unusual for sculpins, the soft sculpin swims around just beneath the surface. These specimens are probably large larvae which were produced the preceding year and which have still not settled permanently to the bottom. This creature's adult and larval anatomy resembles that of snailfishes perhaps as much as sculpins.

Too small and soft.

125 Thornback Sculpin

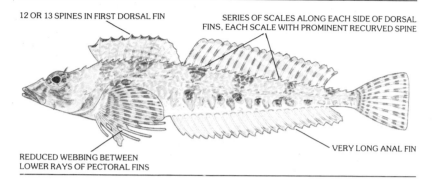

12 OR 13 SPINES IN FIRST DORSAL FIN

SERIES OF SCALES ALONG EACH SIDE OF DORSAL FINS, EACH SCALE WITH PROMINENT RECURVED SPINE

VERY LONG ANAL FIN

REDUCED WEBBING BETWEEN
LOWER RAYS OF PECTORAL FINS

SPECIES: *Paricelinus hopliticus*—from the Greek *para*, *Hicelos* and *hopliticus*, meaning "near," God of Sleep," and "armed."

ALTERNATE NAMES: none.

MAXIMUM RECORDED SIZE: 20 cm (7.8 inches).

DISTRIBUTION: Cortez Bank, off San Diego, southern California, to Banks Island, Hecate Strait, northern British Columbia.

Few anglers fish with small hooks at depths sufficiently deep to place a tempting morsel in front of the tiny-mouthed thornback sculpin. Both fisherman and sculpin find this "arrangement" beneficial.

Very rarely observed by divers, the distinctive thornback sculpin consequently offers a worthy challenge for the neoprene-suited naturalist. Look for this often colourful fish on silt-covered barren, rocky reefs or cliff faces, and along their sandy bases at depths below 20 m (66 feet). Propped by its pectoral and tail fins, it will probably be resting upon the bottom scanning its environment for predator or prey. As more and more knowledgeable divers descend into Pacific Northwest waters, species like the thornback sculpin, now considered rare, could possibly become increasingly well known. As in the case of the once rarely-seen longfin sculpin, SCUBA divers have gathered knowledge and have contributed observations of supposedly rare Pacific Northwest species.

Very infrequently does the unsaleable thornback sculpin find itself swept into the belly of a shrimp trawler's net, along with many other creatures, at depths between 20 and 183 m (66 and 604 feet).

The slender thornback sculpin does not swim in shallow waters accessible to beachcombers or dockside observers.

Too slender for serious eating.

126 Silverspotted Sculpin

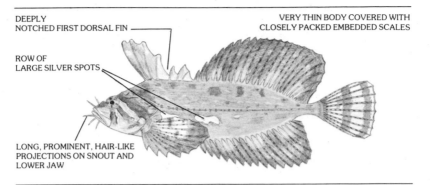

DEEPLY NOTCHED FIRST DORSAL FIN

VERY THIN BODY COVERED WITH CLOSELY PACKED EMBEDDED SCALES

ROW OF LARGE SILVER SPOTS

LONG, PROMINENT, HAIR-LIKE PROJECTIONS ON SNOUT AND LOWER JAW

SPECIES: *Blepsias cirrhosus*—from the Greek *blepo*, meaning "look;" and the Latin *cirrhosus*, "cirrus bearing."

ALTERNATE NAMES: silverspot sculpin, silver spot.

MAXIMUM RECORDED SIZE: 19 cm (7.5 inches).

DISTRIBUTION: San Simeon, northern California, to the Aleutian Islands and Bering Sea, Alaska.

If you are one of the very few anglers who would like to catch the graceful silverspotted sculpin, try float fishing among dense kelp and use live shrimp as bait upon a small hook.

Although the beautiful silverspotted sculpin swims very commonly in shallow, diveable depths, only dedicated aquanauts who persevere through very thick growths of marine plants will likely find it. Its golden or dark green colour, lacy appearance and sinuous swimming behaviour give this large-finned fish the appearance of a ragged piece of kelp wagging in the current. Although recorded from as deep as 37 m (122 feet), it is most abundant at depths less than 10 m (33 feet) and, though elusive, is very accessible to the underwater photographer.

Because commercial harvesters do not consistently fish the shallow kelp forest or intertidal habitats of the Pacific Northwest, the unmarketable silverspotted sculpin very rarely becomes part of their catch.

So well is the golden brown or green silverspotted sculpin camouflaged that the surface-bound naturalist will seldom see it without first using the indispensible dip net. To capture a specimen, scoop through the thick carpet of weed that grows along rocky shores or in tidepools, or hangs from pilings and floats. In late winter mature female silverspotted sculpins deposit blue or light brown eggs upon rocky substrates and by spring these tiny spheres have hatched and ultimately become pale-green and delicate-looking young.

Even a large silverspotted sculpin provides little edible flesh upon its very thin body.

127 Sailfin Sculpin

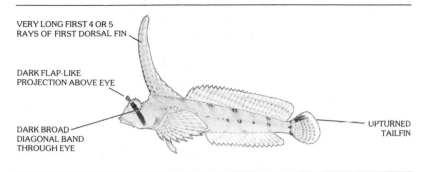

VERY LONG FIRST 4 OR 5
RAYS OF FIRST DORSAL FIN

DARK FLAP-LIKE
PROJECTION ABOVE EYE

DARK BROAD
DIAGONAL BAND
THROUGH EYE

UPTURNED
TAILFIN

SPECIES: *Nautichthys oculofasciatus*—from the Greek *nautys* and *ichthys*, meaning "sailor" and "fish;" as well as the Latin *oculus* and *fasciatus*, "eyes" and "banded."

ALTERNATE NAMES: sailorfish, sailor fish.

MAXIMUM RECORDED SIZE: 20 cm (8 inches).

DISTRIBUTION: San Miguel Islands, southern California, to St. Lawrence Island, in the Bering Sea, Alaska.

Occasionally at night a dockside angler might catch the tiny-mouthed sailfin sculpin using small live shrimp as bait.

A common sight for the night diver. The graceful whitish or brownish sailfin sculpin swims over the bottom by holding its body stiff while moving its dorsal and anal fin rays independently, one after another in a series of graceful, continuous waves. Although most abundant in shallow waters where it cruises along rocky outcroppings, along cliff faces or upon adjacent sandy areas, this sinuous swimmer may live around jetties and pilings. Patiently watch a specimen lower its long first dorsal rays over its head, then move them back and forth in front of itself—a distinctive behaviour often occurring just before this sculpin gulps a tiny shrimp or other prey. By day, the nocturnally active sailfin sculpin hides inside caves or crevices and usually beyond view. In spring female sailfin sculpins lay small clusters of adhesive orange eggs among mussels in the upper intertidal zone, and the tiny larvae which hatch from these eggs, sporting exaggerated, enlarged fins, spend their first few weeks at the surface.

The resourceful sailfin sculpin, sometimes pinkish or orange at great depths, may invade prawn traps down to 110 m (363 feet). Shrimp trawlers, too, sometimes drag up this unsaleable species.

Watch for the slow swimming sailfin sculpin at night from well-lighted docks or jetties. Receding tides occasionally trap it in cool northern tidepools.

Too thin and slender for serious eating.

128 Grunt Sculpin

LARGE HEAD WITH
HEAVY BLUNT RIDGES

SHORT STOUT BODY

LONG SNOUT

LOWER RAYS OF
PECTORAL FINS WITH
VIRTUALLY NO
WEBBING

SPECIES: *Rhamphocottus richardsoni*—from the Greek *rhampho* and *cottus*, meaning "snout" and "sculpin;" and honouring explorer-naturalist John Richardson.

ALTERNATE NAMES: Richardson's sculpin, gruntfish, pigfish, northern sea horse.

MAXIMUM RECORDED SIZE: 8.3 cm (3.3 inches).

DISTRIBUTION: Santa Barbara, southern California, to the Bering Sea coast of Alaska, and one possible record from Honshu, Japan.

Too small to capture on conventional angling gear, the tiny-mouthed grunt sculpin would not even be good bait.

Finding a grunt sculpin, with its big head, pig-like snout and orange tail fin is a highlight of any dive. Search for it along rocky shores or reefs, particularly among encrusting invertebrate animals growing there. Often the snout protrudes from an empty giant barnacle casing and makes this casing appear to contain a live but closed barnacle within. This tiny fish may turn around within the casing, stick its tail fin out and wave it back and forth, and appear to be a feeding barnacle actively waving its cirri to capture microscopic planktonic prey. If such natural shelter is unavailable, the irrepressible grunt sculpin often colonizes debris, such as discarded bottles or cans. When moving along the sea floor this creature rarely swims but instead "walks" along on its finger-like lower pectoral fin rays. Any diver can catch a specimen by hand and actually feel it producing quiet vibrations which, perhaps, suggest its name.

More frequently wandering into prawn traps than being incidentally scooped up by shrimp trawls at depths to 200 m (660 feet), the tiny grunt sculpin has no market value.

Very rarely trapped in tidepools, the amusing grunt sculpin does seek shelter under floats or piers. The aggressive, active females often claim a bottle or can when such a container houses a potential mate.

How could anyone think of eating this cute little fellow?

162

129 Cabezon

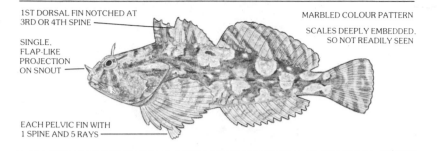

1ST DORSAL FIN NOTCHED AT 3RD OR 4TH SPINE

SINGLE, FLAP-LIKE PROJECTION ON SNOUT

MARBLED COLOUR PATTERN

SCALES DEEPLY EMBEDDED, SO NOT READILY SEEN

EACH PELVIC FIN WITH 1 SPINE AND 5 RAYS

SPECIES: *Scorpaenichthys marmoratus*—from the Greek *scorpaena* and *ichthys*, meaning "fish with a poisonous sting" and "fish;" and the Latin *marmoratus*, "marbled."

ALTERNATE NAMES: giant marbled sculpin, giant sculpin, blue garnet, bull cod*, bullhead*.

MAXIMUM RECORDED SIZE: 99 cm (39 inches) and 14 kg (30 pounds)

DISTRIBUTION: Punta Abrejos, central Baja California, Mexico, to Sitka, southeastern Alaska.

Especially popular in California waters where it ranks among the top ten inshore sportfish, the bottom dwelling cabezon ravenously seizes such baits as clams, marine worms, shrimps, crabs and small fishes. Cast for this large sculpin from jetties, boats or shorelines, particularly near kelp beds but expect only one strong short burst of action before landing it.

Often a diver is startled when literally bumping into the well-camouflaged cabezon before it bolts powerfully for nearby kelp or other cover. However, in late winter adult male cabezon do not bolt so readily but steadfastly guard large clumps of adhesive wine-red and purple eggs deposited on rocky out croppings or sunken logs by mates which have already departed. Impressed by its size, many spearfishermen take the sedentary cabezon.

Californian setliners enthusiastically harvest the popular cabezon along with other species, but Pacific Northwest commercial fishermen still underutilize it.

After just a month of life, the silvery almost square-looking young cabezon, very different from its parents, may hide among the floating weeds associated with summer tidelines. Tidepools frequently trap older more adult-looking specimens. Often living in very shallow water, a large adult cabezon may lurk among marine plants within centimeters of the surface but because of its well-camouflaged form, it is difficult to see as it lies still.

Excellent—even occasional specimens tinged with green lose this colour upon being cooked. Try it barbecued. *Do not eat the roe:* it is highly toxic.

*—incorrect

163

130 Great Sculpin

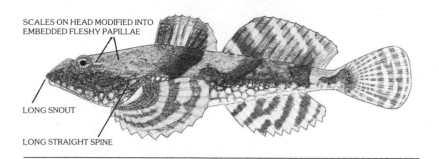

SCALES ON HEAD MODIFIED INTO
EMBEDDED FLESHY PAPILLAE

LONG SNOUT

LONG STRAIGHT SPINE

SPECIES: *Myoxocephalus polyacanthocephalus*—from the Greek *myoxos*, *cephalos*, *poly* and *acantha*, meaning "door mouse," "head," "many" and "spine."

ALTERNATE NAMES: bullhead*.

MAXIMUM RECORDED SIZE: 76 cm (30 inches).

DISTRIBUTION: Puget Sound, northern Washington, to the Bering Sea coast of Alaska, through the Aleutian chain, to Kamchatka, U.S.S.R., and Hokkaido, northern Japan.

 The sluggish great sculpin greedily gulps down baits of nearly any size or kind lowered onto the shallow bottom. Jetty or wharf-bound anglers most often take it, sometimes with a very slowly retrieved lure. Unpopular with most hook and liners, this sculpin provides little or no sport and interferes with the pursuit of other more desirable quarry.

 Divers sometimes encounter the grayish great sculpin resting upon silt-covered rocky outcroppings or adjacent sandy substrates at depths less than 20 m (66 feet). Wharves, pilings and their associated underwater debris often shelter this large sculpin. Although its usually drab coloration matches well with its generally dull surroundings, some more attractively hued specimens lurk among more colourful encrusting plants and animals.

 Trawlers harvesting important groundfish occasionally find an unwanted great sculpin in the catch while set liners sometimes disdainfully remove one or two from the gear.

 Sometimes dockside naturalists may see the large-headed gray form lying motionless on the bottom, often around pilings. More likely, though, beachcombers strolling along sandy shores in spring could notice tiny, barely recognizable juvenile specimens hiding in very shallow water, perhaps amid eelgrass.

Make an unusual fish loaf. Melt butter, stir in flour, some milk, then cook and cool before adding the great sculpin, celery, bread crumbs, parsley, onion, salt and tomato sauce. Finally mold in a pan and bake.

*—incorrect

131 Buffalo Sculpin

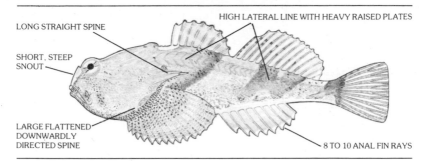

LONG STRAIGHT SPINE

HIGH LATERAL LINE WITH HEAVY RAISED PLATES

SHORT, STEEP SNOUT

LARGE FLATTENED DOWNWARDLY DIRECTED SPINE

8 TO 10 ANAL FIN RAYS

SPECIES: *Enophrys bison*—from the Greek *en*, *ephrys* and *bison*, meaning "on," "eyebrow," and "bison."

ALTERNATE NAMES: buffalo fish, bullhead*.

MAXIMUM RECORDED SIZE: 37 cm (14.6 inches).

DISTRIBUTION: Monterey, central California, to Kodiak Island, Gulf of Alaska.

While trying to catch flounders, greenlings and rockfish, shorebound anglers often take the steep-snouted buffalo sculpin. It ravenously gulps such baits as marine worms, clams, shrimps, mussels and fish lowered onto weedy, boulder-strewn shores, or under floats and piers. After seizing the hook and meekly tugging the line, this lethargic sculpin usually disappoints the angler. Handle this fish carefully: avoid its large sharp preopercular spines, but notice the hum that emanates from the vibrating and annoyed specimen.

Motionless and concealed amid plant or animal-encrusted rocks at depths less than 15 m (50 feet), the variably coloured buffalo sculpin is well camouflaged. This spotted, mottled and/or blotched fish often remains very still and a careful, deliberate diver can sometimes actually pick one up. During February or March, mature buffalo sculpins guard large clumps of amber-coloured eggs and thus provide the underwater photographer with an especially good opportunity "to shoot." After hatching and spending a brief interlude at the surface, in late spring the barely recognizable young descend to populate shallow eelgrass-choked beaches.

Commercial fishermen rarely catch the unmarketable buffalo sculpin because they are not looking for it, have not the proper gear, and do not fish in shallow enough water.

Only occasionally found in tidepools, the shallow-water buffalo sculpin may be visible from a float or pier. More visible young specimens feed upon many smaller creatures, and even on sea lettuce.

The small amount of flesh available from even a large buffalo sculpin requires some seasoning and is perhaps suitable only as an ingredient in a dip.

*—incorrect

165

132 Red Irish Lord

SINGLE DORSAL FIN NOTCHED TO FORM 3 STEPS

NOSTRILS WITHOUT FLAPS

BAND OF SCALES, NOT MORE THAN
4 OR 5 SCALES WIDE, SURROUNDING DORSAL FIN

SPECIES: *Hemilepidotus hemilepidotus*—from the Greek *hemi* and *lepidotus*, meaning "half" and "scaled."

ALTERNATE NAMES: red irish lord sculpin, spotted irish lord, red sculpin, bullhead*.

MAXIMUM RECORDED SIZE: 51 cm (20 inches).

DISTRIBUTION: Monterey Bay, central California, to the Pribiloff Islands, Bering Sea, Alaska.

A gluttonous feeder that prefers crabs, shrimp, barnacles, mussels and small fishes, the red irish lord usually swallows the bait completely. Although a fairly common, easily landed catch for bottomfishermen angling along rocky shorelines, from boats over shallow reefs or from piers, this sluggish fish rates low with most sportsmen.

Popular with underwater photographers because of its variable, often brilliant coloration and sedentary behaviour, the red irish lord flourishes in shallow, easily-dived habitats. So well does this common creature blend with its usual background of colourful encrusting plant and animal life, the filmmaker may have trouble highlighting it in order to make it "stand out" in the photo; with a little careful prodding or even gentle handling, though, the photographer may relocate a co-operative specimen to a more suitable background. Spawning occurs in winter when ripe females deposit large masses of pink, yellow, purple, or blue eggs, which the easily-approached and attentive male tenaciously guards.

When harvesting valuable groundfish, set liners working over precipitous bottoms at times incidentally take some very colourful red irish lords but trawlers fishing over flat silty habitats drag up the more drably hued. Because profits do not result from the red irish lord, which is nearly all head and fin, fishermen disdainfully discard them.

Common inhabitants of the wharf and piling habitat, adult red irish lords seldom languish in tidepools where younger specimens frequently hide.

Cook, then flake some red irish lord fillets; mix with onion, green pepper, chopped celery, salt, crackers and mayonnaise. Chill the mixture and stuff it in hollowed tomatoes and serve on lettuce.

*—incorrect

166

133 Brown Irish Lord

SINGLE DORSAL FIN DEEPLY NOTCHED TO FORM 3 STEPS

NOSTRILS WITH FLAPS

BAND OF SCALES, 6 TO 8 SCALES WIDE
SURROUNDING DORSAL FIN

SPECIES: *Hemilepidotus spinosus*—from the Greek *hemi* and *lepidotus*, meaning "half" and "scaled;" and the Latin *spinosus*, "spiny."

ALTERNATE NAMES: bullhead*.

MAXIMUM RECORDED SIZE: 29 cm (11.3 inches).

DISTRIBUTION: Santa Barbara Island, southern California, to Puffin Island, southeastern Alaska.

Perhaps the hardy angler who baitfishes along the rugged, surf-swept outer coast might inadvertently catch the sluggish brown irish lord. An inhabitant of rocky bottoms at depths to 97 m (318 feet), this secretive fish provides little sport anyway.

Only divers who brave the surge and poor visibility of rocky shallows bordering directly on the Pacific Ocean might happen upon the usually motionless brown irish lord. Nestling among the cracks and crevices of solid outcroppings, particularly where pink, branching coralline algae coats the bottom, this well camouflaged creature is most difficult to find. It remains still, however, allowing for prolonged viewing, or even allows itself to be moved by a careful underwater photographer. Various environmental factors—the surge and poor visibility—combined with the brown irish lord's cryptic coloration often foil the underwater photographer.

Incidental captures of the brown irish lord very rarely occur because commercial harvesters seldom set their gear in the shallow, rocky environment of this creature.

Because this inconspicuous creature huddles among the weed-covered rocks of outer coast tidepools or adjacent shorelines, a beachcomber may not even know a brown irish lord is present unless scooping it up in a dip net.

Why not try brown irish lord taipei? Gently fry onion until tender; add cooked rice and soy sauce. Pour in beaten eggs. Cook while stirring gently before finally adding water chestnuts and the cooked, flaked fillets.

*—incorrect

134 Pacific Staghorn Sculpin

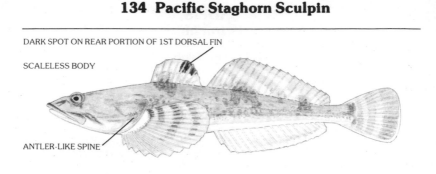

DARK SPOT ON REAR PORTION OF 1ST DORSAL FIN

SCALELESS BODY

ANTLER-LIKE SPINE

SPECIES: *Leptocottus armatus*—from the Greek *leptos* and *cottus*, meaning "slender" and "sculpin;" and the Latin *armatus*, "armed."

ALTERNATE NAMES: staghorn sculpin, cabezon*, bullhead*.

MAXIMUM RECORDED SIZE: 46 cm (18 inches).

DISTRIBUTION: Bahia San Quintin, northern Baja California, Mexico, to Izembek Lagoon, on the southeastern Bering Sea coast of Alaska.

A very common catch, particularly for young anglers, the Pacific staghorn sculpin flourishes in bays and inlets with gently sloping silt or mud bottoms. Despised as a bait-stealing nuisance by most fishermen, it greedily devours virtually any bait—even its own kind—but after giving one or two meek tugs, even a large one easily comes ashore. Use caution when removing the hook from a squirming specimen because its sharp, ragged preopercular spines can inflict a painful wound.

Because the Pacific staghorn sculpin often buries itself in silty, muddy bottoms, a diver must look closely to see its grayish, olive-green form. Carefully "excavate" a specimen, then watch it wiggle its body and scoop with its pectoral fins to re-bury itself. An abundance of young staghorn sculpins live under floats and especially in the brackish waters of river estuaries where various sea birds feed upon them.

Trawlers straying over shallow muddy substrates inadvertently take the unmarketable Pacific staghorn sculpin from depths less than 91 m (300 feet).

Docks or jetties built over gently sloping bays provide excellent viewing platforms from which to watch the slow-swimming Pacific staghorn sculpin periodically move about in search of prey; however, summer bathers shuffling through tidal flats often startle juveniles which dart away in all directions.

The soft texture and bland flavour of the Pacific staghorn sculpin make it suitable for a fish paste and little else.

*—incorrect

168

THE TOADFISHES
(Family: Batrachoididae)

The moderate-sized Batrachoididae, which translates from the Greek to mean "frog-like family," possesses about 55 recognized living species of midshipmen and toadfishes. Nearly all of the sluggish, bottom-dwelling toadfishes flourish in tropic latitudes and the species are spread fairly evenly around the globe. Only a very few members of this primarily marine family invade brackish or fresh water realms while the remainder frequent the coastal environment. Whether members of this family live in dark, cold, deep water habitats or thrive in shallow bays, they are generally found upon soft muddy and sandy bottoms. Much of the time batrachoidid fishes including the single Pacific Northwest species, the plainfin midshipman, bury themselves in the silt while awaiting prey or hiding from predators, but also rise off the bottom to move about.

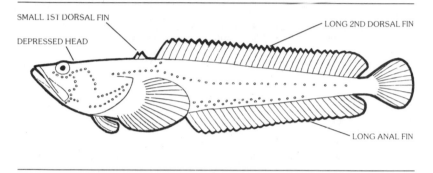

SMALL 1ST DORSAL FIN

LONG 2ND DORSAL FIN

DEPRESSED HEAD

LONG ANAL FIN

It is popularly believed that the underwater world is silent, but such is not the case. The fascinating toadfishes are among the noisiest of aquatic animals: nearly all produce a variety of sounds by vibrating certain inner ear bones against connecting gas-filled swim bladders to achieve a resonating effect. Long playing phonograph records document an audible repertoire of grunts, squeaks, whistles, hums and croaks—sounds which allow these fishes to communicate territorial and breeding information to one another.

That recognizable toadfishes appeared on earth at least 12 million years ago is documented through batrachoidid fossils unearthed from Miocene sedimentary rocks. A small, distinctive group, toadfishes apparently have no really close surviving relatives, with the possible exception of the superficially different clingfishes. Only a few scientists and aquarium enthusiasts find toadfishes of great interest because most varieties attain the modest maximum size of 51 cm (20 inches) or less.

135 Plainfin Midshipman

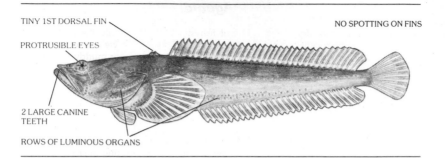

TINY 1ST DORSAL FIN

PROTRUSIBLE EYES

2 LARGE CANINE TEETH

ROWS OF LUMINOUS ORGANS

NO SPOTTING ON FINS

SPECIES: *Porichthys notatus*—from the Greek *porous* and *ichthys*, meaning "pore" and "fish;" and the Latin *notatus*, "spotted."

ALTERNATE NAMES: northern midshipman, midshipman, singing fish, bullhead*, cabezone*.

MAXIMUM RECORDED SIZE: 38 cm (15 inches).

DISTRIBUTION: Angel de la Guardia, in the Gulf of California, around the tip of Baja California, Mexico, and north to Sitka, southeastern Alaska.

Because the sedentary plainfin midshipman, with its upturned and tooth-studded jaws, rests very still on a muddy or silty bottom, to catch one an angler would have to drop the bait virtually on top of the creature's mouth.

Particularly during daylight, the abundant plainfin midshipman remains buried in sand or mud and divers therefore seldom sight it. However, at night this darkly coloured fish commonly emerges from the mud and often hovers nearly motionless just off the bottom. Look closely at its undersides for the characteristic rows of bioluminescent photophores, silvery little dots which form a pattern like the buttons that decorate the jacket of a midshipman. Studies show that these photophores produce light created from a substance called luciferin, which is also present in one of the shrimp-like species that the plainfin midshipman preys upon.

Of no direct commercial value, the despised plainfin midshipman, with its sharp needle-like teeth catches on and often fills fine-meshed shrimp trawls dragged from depths to 366 m (1,200 feet).

During a summer low tide, search along high and dry muddy or sandy shores and turn over rocks to see a male plainfin midshipman guarding a nest. On the underside of the overturned stone, notice the yellow, kernel shaped eggs or even the tiny attached young. Carefully replace the rock to hide the nest and its guardian.

Though tasty, even the largest plainfin midshipman possesses little edible flesh.

*—incorrect

170

THE POACHERS
(Family: Agonidae)

The Agonidae, a moderately large group of fishes, includes not only the starsnouts, alligatorfishes and the rockhead, but also the poachers. These fishes live in a wide variety of habitats ranging from shallow rocky and intertidal shores to sandy or muddy substrates at considerable depths. Of the approximately 50 known varieties, most species live in the northern Pacific Ocean, at least 20 along North America's western shoreline, and a comparable assortment including the largest at 40 cm (16 inches) in Asian Pacific waters. The Arctic and northern Atlantic Oceans provide coastal habitats for other members of this group.

Primarily because of their rigid or relatively inflexible bodies, the small and bone-plated agonids swim with a sculling motion of their pectoral fins while dragging their bodies like rudders. However, during times of extreme danger or stress, they adopt a frenzied wriggling as an escape mechanism—especially the numerous elongated species. The few short and stubby poachers crawl along, using their pectoral, anal and caudal fins.

BODY COMPLETELY COVERED WITH ROWS OF BONY PLATES
WHICH MEET BUT DON'T OVERLAP

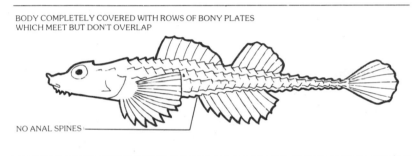

NO ANAL SPINES

Although few fossilized remains have been found, paleontologists have determined that ancestors of the present-day—but essentially ignored—poachers lived in shallow marine seas at least 40 million years ago, during the Eocene era. The poachers' closest modern relatives are the other mailcheeked fishes, the sculpins, rockfishes, greenlings, snailfishes, and sablefishes. All have a common feature: a large heavy bone, known as a "suborbital stay," beneath each eye.

This book includes 15 poachers of the Pacific Northwest, but does not include others which are not commonly encountered: the pricklebreasted poacher—a thin elongate species with many hair-like prickles on its breast just behind its gills; the warty poacher—an elongate species with a long spineless head; the pixie poacher—which may be only the young of another species, the bluespotted poacher—a fish similar to the one on page 179, but without a black edge on its first dorsal fin and with two barbels at each corner of its mouth; the smootheye poacher—a fish similar to the one on page 179 but with no spiny scales on its eyes; and the fourhorn poacher—a small stubby form with a steep forehead, high set eyes and large spines on the top of its head.

136 Northern Spearnose Poacher

POINTED SNOUT WITH 2 BLUNT
SKIN COVERED SPINES

MOUTH ON
UNDERSIDE

LIGHT SPOT IN CENTRE
SURROUNDED
BY DARK AREA

SPECIES: _Agonopsis vulsa_—from the Greek _agon_ and _opsis_, meaning "without joints" and "like" and the Latin _vuls_, "smooth."

ALTERNATE NAMES: windowtail poacher, window-tailed sea-poacher, dark alligatorfish, northern spearnose.

MAXIMUM RECORDED SIZE: 20 cm (8 inches).

DISTRIBUTION: Point Loma, southern California, to southeastern Alaska.

Only very rarely might a bottom-fishing angler, using very tiny hooks baited with a bit of clam or shrimp, catch the slender, heavily armoured northern spearnose poacher. Its stiff, spiny body makes a poor bait.

Nocturnal searchers diving along sandy bottoms adjacent to rocky outcroppings and cliff faces at depths below 10 m (33 feet) frequently come across the northern spearnose poacher. Watch and perhaps see one grovelling through the sand to find a meal of tiny worms or clams. When it moves, this grayish-brown creature propels itself with its large pectoral fins and drags its stiff body rudder-like behind it. However, it usually rests motionless on the bottom and its colour makes it difficult to see. Careful and resourceful underwater photographers can actually pick up a specimen and place it on a rock to get a better "shot" against a contrasting background.

Fine-meshed shrimp trawls towed along soft level bottoms frequently take the unsaleable northern spearnose poacher as an incidental, at depths between 10 and 180 m (33 and 594 feet).

During spring and summer the juvenile northern spearnose poacher commonly swims toward lights shining upon the water from piers or jetties. Initially unrecognizable because of their proportionately large dark tails and the huge spines located between their eyes, these young specimens often feed upon tiny, free-swimming crab larvae.

Removing the heavy armour from even a large northern spearnose poacher is hardly worth a chef's time.

137 Sturgeon Poacher

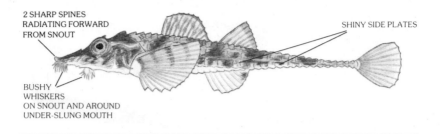

2 SHARP SPINES
RADIATING FORWARD
FROM SNOUT

SHINY SIDE PLATES

BUSHY
WHISKERS
ON SNOUT AND AROUND
UNDER-SLUNG MOUTH

SPECIES: *Agonus acipenserinus*—from the Greek *agonus*, *acipenser* and *inus*, meaning "without joints," "sturgeon," and "like."

ALTERNATE NAMES: sturgeon-like sea-poacher.

MAXIMUM RECORDED SIZE: 30 cm (12 inches).

DISTRIBUTION: Eureka, northern California, to Barrow, Alaska; through the Aleutian chain to the Komandorski Islands and the Anadyr Gulf, USSR.

Because the tiny-mouthed sturgeon poacher could only take a very minute hook, most anglers never see this creature. Very rarely do larger hooks meant for other quarry snag this species.

Look for the yellowish sturgeon poacher while cruising along soft, nearly level bottoms at depths from the surface among eelgrass beds to 30 m (100 feet). Before consuming bottom-dwelling invertebrates such as shrimps and other shrimp-like creatures, this slow-moving fish must first locate them by grovelling through the sand or silt with its bushy, tastebud-bearing cirri, prominently located beneath its mouth. In the USSR the dark sturgeon poacher spawns in spring, but as yet no documented information exists about breeding in North American populations.

While seeking shrimps, inshore trawlers incidentally scoop up the unsaleable sturgeon poacher from flat bottoms at depths to 60 m (200 feet).

Wade out into the shallows of a muddy-bottomed bay and look for the yellowish-brown sturgeon poacher as it huddles among eelgrass or flees, startled by your threatening steps. It usually swims so slowly that a naturalist can catch it by dipnet or even by hand. If the water is clear, a sharp-eyed dockside naturalist, too, might notice the well-camouflaged sturgeon poacher below the pier.

It is not worthwhile attempting to remove the modest amount of flesh from the spiny hard body of even a large sturgeon poacher.

138 Pygmy Poacher

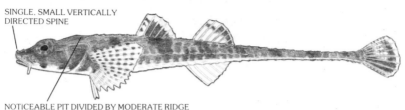

SINGLE, SMALL VERTICALLY DIRECTED SPINE

NOTICEABLE PIT DIVIDED BY MODERATE RIDGE

SPECIES: *Odontopyxis trispinosa*—from the Greek *odons* and *pyxis*, meaning "tooth," and "box," and the Latin *tres* and *spinus*, "three" and "spine."

ALTERNATE NAMES: pygmy sea-poacher.

MAXIMUM RECORDED SIZE: 9.5 cm (3.8 inches).

DISTRIBUTION: Isla Cedros, central Baja California, Mexico, to southeastern Alaska.

 Anglers are unlikely to find the grayish pygmy poacher dangling from their lines because it is far too small to be hooked on conventional fishing tackle.

 Very observant divers may frequently encounter the pygmy poacher resting upon a sandy or muddy bottom at depths below 5 m (16 feet). Sightings are possible by day or night, but unless this well-camouflaged creature moves, it is nearly impossible to see at any time. If patient, you may even see one approach another, twist its body into a crescent shape directly in front of the second individual and wave its pectoral fins—a social behaviour as yet unexplained. A tiny subject that moves just as the exasperated underwater photographer is about to "shoot," the pygmy poacher blends in almost perfectly with its background and is a truly challenging subject for the diving camera buff. Divers exploring waters within the northern part of the pygmy poacher's range are more likely to see this fish because in the southern areas it lives at much greater depths—300 m (990 feet) or more.

Although the shrimp fisherman may incidentally take the pygmy poacher in his fine-meshed trawls, it is of no commercial value.

 A watchful pierside stroller may perhaps notice a pale gray pygmy poacher, attracted by lights, struggling toward the surface at night.

 The tiny, hard carcass of a pygmy poacher is most unappetizing to the gourmet.

174

139 Spinycheek Starsnout

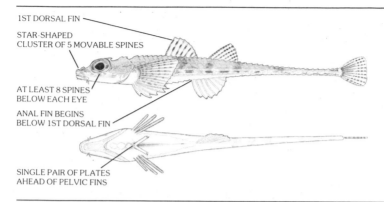

1ST DORSAL FIN

STAR-SHAPED
CLUSTER OF 5 MOVABLE SPINES

AT LEAST 8 SPINES
BELOW EACH EYE

ANAL FIN BEGINS
BELOW 1ST DORSAL FIN

SINGLE PAIR OF PLATES
AHEAD OF PELVIC FINS

SPECIES: *Bathyagonus infraspinata*—from the Greek *bathy* and *agonus*, meaning "deep" and "without joints," and the Latin *infra* and *spinata*, "below" and "spined."

ALTERNATE NAMES: spinycheek starnose, spiny-cheeked star-snout.

MAXIMUM RECORDED SIZE: 12.1 cm (4.8 inches).

DISTRIBUTION: Eureka, northern California, to the Bering Sea coast of Alaska.

Too tiny to be caught with conventional angling gear, the hard-bodied spinycheek starsnout is poor bait.

Although not rare and an inhabitant of depths accessible to SCUBA enthusiasts, few divers actually ever see the inconspicuous spinycheek starsnout. It lives upon level sandy or muddy substrates which appear barren of life, and only when this slender grayish fish moves does it become noticeable. Upon actually finding a spinycheek starsnout, the sharp-eyed aquanaut usually gets a good long look because a specimen seldom moves very far or very fast.

While not specifically seeking the tiny and unmarketable spinycheek starsnout, commercial shrimp harvesters nevertheless take it in their fine-meshed nets towed over flat, soft bottoms at depths between 20 and 200 m (66 and 660 feet), and then discard it. The indirect value of this bony-plated fish as forage for valuable groundfish such as codfishes, flounders, and rockfishes is difficult to assess.

A resident of deep environs, inaccessible to surface-bound naturalists, the pale spinycheek starsnout may best be seen in a public aquarium. In an aquarium tank an active specimen sometimes bends its stiff body from side to side and then waves its pectoral fins at any nearby conspecific, one of its own kind. Whether this is a behaviour common in its natural environment is as yet unknown.

Too small and hard to eat.

140 Gray Starsnout

STAR-SHAPED CLUSTER OF
5 MOVABLE SPINES

DORSAL FINS

ANAL FIN BEGINS
UNDER SPACE BETWEEN
DORSAL FINS

SINGLE PAIR OF PLATES
AHEAD OF PELVIC FINS

SPECIES: ***Bathyagonus alascana***—from the Greek *bathy* and *agonus*, meaning "deep" and "without joints," and emphasizing Alaska.

ALTERNATE NAMES: gray star-snout, gray starnose.

MAXIMUM RECORDED SIZE: 13 cm (5 inches).

DISTRIBUTION: the California-Oregon border to the Bering Sea coast of Alaska.

A small, slow-moving creature that does not readily attract the interest of larger, desirable sportfish, the stiff-bodied gray starsnout would make poor bottomfish bait.

Only very observant divers who make forays along sandy or muddy bottoms at depths greater than 20 m (66 feet) might perhaps find the inconspicuous gray starsnout resting propped upon its pectoral and tail fins. At other times it is either belly down on the substrate or swimming in a number of short bursts by sculling with its large pectoral fins and dragging its slender, stiff body behind. When photographing a specimen, the thoughtful underwater camera buff approaches the wary gray starsnout from "down current" so that any stirred up silt or debris does not drift over the specimen and spoil a good "shot." A steady-handed and patient diver can capture an unstartled specimen with a careful grasp and then study more closely its armour-encased body. Its diet consists primarily of invertebrate animals such as shrimps and worms, which it finds by persistent grovelling along the soft sea floor.

Though unmarketable, the insignificant gray starsnout nevertheless enters the catch of commercial shrimp trawlers who tow fine-meshed nets over flat, soft bottoms at depths between 20 and 250 m (66 and 825 feet).

Any beachcombers and pierside naturalists who have made intertidal or shallow-water sightings of the slender gray starsnout have apparently not recorded them.

Too tiny, spiny and hard.

141 Bigeye Poacher

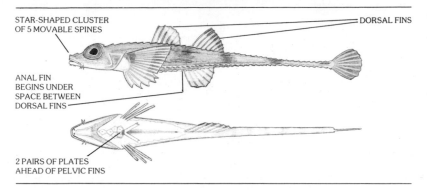

STAR-SHAPED CLUSTER
OF 5 MOVABLE SPINES

DORSAL FINS

ANAL FIN
BEGINS UNDER
SPACE BETWEEN
DORSAL FINS

2 PAIRS OF PLATES
AHEAD OF PELVIC FINS

SPECIES: ***Bathyagonus pentacanthus***—from the Greek *bathy*, *agonus*, *penta* and *acanthus*, meaning "deep," "without joints," "five" and "spine."

ALTERNATE NAMES: bigeye starsnout, bigeye starnose.

MAXIMUM RECORDED SIZE: 23 cm (9.3 inches).

DISTRIBUTION: Cortez Bank, near San Diego, southern California, to the Chirikof Islands in the Gulf of Alaska.

Very few anglers would encounter the grayish bigeye poacher because it lives in deep water below 100 m (330 feet), where few bait-fishermen lower hooks small enough to catch it. This fish would not even make good bait.

Documented only from depths well below the range that the average SCUBA diver would patrol, the slender bigeye poacher lives upon darkened, gently sloping sea floors of mud or sand. However, someone might amend established records and often a knowledgeable aquanaut leads the way with initial sightings of such seldom-seen species.

A shrimp trawler occasionally scoops up the bigeye poacher while towing a fine-meshed net along level sea floors. Actually, Joe Bauer of Steveston documented the first specimen of this pale poacher ever taken in British Columbia in just this manner. Mr. Bauer, a commercial fisherman and knowledgeable naturalist, noticed it while carefully sorting through a large catch hauled from a sandy bottom at 125 m (413 feet) off Galiano Island and reported it to Dr. Wilbert Clemens, who was updating *Fishes of the Pacific Coast of Canada* in 1961.

That a surface-bound naturalist would ever encounter the slow-swimming bigeye poacher anywhere is extremely unlikely, except in a large public aquarium.

Far too slender and bony for serious dining.

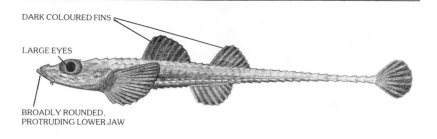

DARK COLOURED FINS

LARGE EYES

BROADLY ROUNDED,
PROTRUDING LOWER JAW

SPECIES: *Bathyagonus nigripinnis*—from the Greek *bathy* and *agonus*, meaning "deep" and "without joints;" and the Latin *nigri* and *pinnis*, "black" and "fin."

ALTERNATE NAMES: black-finned sea-poacher, blackfin starsnout, blackfin starnose.

MAXIMUM RECORDED SIZE: 21 cm (8.5 inches).

DISTRIBUTION: Eureka, northern California, to the Bering Sea coast of Alaska, through the Aleutian chain to the Komandorski Islands and Avachin Bay, USSR.

Very few anglers encounter the thin blackfin poacher because they rarely lower hooks deep enough that would be tiny enough to be effective in capturing such a small-mouthed fish.

The elongate blackfin poacher thrives only at depths well below those explored by SCUBA aficionados. Somewhere within its known distribution though, it may frequent a locale much shallower than 100 m (330 feet) and may possibly be observed by a diver. With luck he or she will be an underwater photographer who obtains good film evidence to identify the specimen positively. Often the beautiful blue of the pectoral fins contrasts sharply with the brownish gray of the body, and would make a striking photo for the diver's den.

While the small, stiff blackfin poacher is of no direct monetary value, shrimp trawlers frequently net specimens as unwanted and quickly discarded catches. Flourishing on fine sand or mud bottoms as far down as 1,250 m (4,125 feet), this large-eyed creature ranks as one of the deepest dwelling members of the poacher family.

Public aquaria infrequently receive live blackfin poachers from interested and conscientious commercial fishermen who manage to keep these little fish alive in a bucket of cold, periodically changed, sea water . Only through such laudable efforts may most people observe this fish.

Too slender and hard.

143 Blacktip Poacher

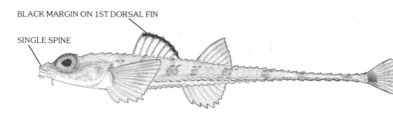

BLACK MARGIN ON 1ST DORSAL FIN

SINGLE SPINE

SPECIES: *Xeneretmus latifrons*—from the Greek *xen* and *eretmus*, meaning "strange" and "oar;" and from the Latin *latis* and *frons*, "wide" and "forehead."

ALTERNATE NAMES: blackedge poacher, black-tipped sea-poacher.

MAXIMUM RECORDED SIZE: 19 cm (7.5 inches).

DISTRIBUTION: Cabo Colnett, northern Baja California, Mexico, to Ucluelet, Vancouver Island and Sechelt, southern British Columbia.

Anglers almost never make contact with the tiny-mouthed blacktip poacher because of the creature's small size.

While enjoying a dive on a dull overcast day or even at night, the aquanaut may perhaps notice a pale gray blacktip poacher resting upon its pectoral fins at depths of 20 m (66 feet) or more. The northern section of the blacktip poacher's distribution affords the diver the best locales for viewing this species because it lives in much deeper water—over 400 m (1,300 feet)—in southern regions. Search particularly over level or gently sloping bottoms where log booming activities litter the sand or silt with wood chips. When hatched in late spring, a tiny black blacktip poacher larva swims weakly at the surface where it feeds on smaller planktonic "companions." When it settles to the bottom as a metamorphosed survivor which now closely resembles its parents, the young poacher feeds upon shrimps and their kin while at the same time avoids hungry predators such as the Pacific hake, flounders and codfishes.

Although unsought, the common blacktip poacher frequently wriggles among the valuable shrimps dumped aboard trawlers from nets towed over soft substrates.

The dull gray blacktip poacher lives upon the dull gray sea floor well below the surface and is never seen by any non-diving naturalist.

The Pacific Northwest's marine environment contains many delicious, desirable delicacies of seafood—but the small, hard blacktip poacher is not one of them.

179

144 Tubenose Poacher

THICK PROMINENT FORWARDLY-
DIRECTED WHISKER
ON LOWER JAW

2 DORSAL FINS

GILL MEMBRANES UNITED AT THROAT

SMOOTH PLATES

ELONGATE BODY

SPECIES: *Pallasina barbata*—honouring early naturalist Petrus Pallas, and from the Latin *barbata*, meaning "bearded."

ALTERNATE NAMES: tubesnout poacher.

MAXIMUM RECORDED SIZE: 14 cm (5.8 inches).

DISTRIBUTION: Bodega Bay, Mendocino County, northern California, to Avatcha Bay, Kamchatka Peninsula, south western Alaska (another race lives off the Pacific coast of the USSR.)

Unless an angler were unintentionally to snag or foul-hook a tubenose poacher, this tiny-mouthed fish is impossible to catch. It is a poor bait anyway.

Divers venturing into shallows might observe the dark green tubenose poacher either where eelgrass grows in shallow, sheltered sandy coves or where surf grass clings tightly to rocky depressions along surge swept shorelines. Look particularly among the fronds and floats of the dense kelp plants: the tubenose is one of the few poachers that consistently and actively swims well off the bottom.

Only occasionally do large tubenose poachers invade deep, level-bottomed habitats where fine-meshed trawl nets can scoop them up. Fishermen discard it because it is far too small for market.

Large tidepools or rocky shorelines choked with marine plants are prime locales for the elongate tubenose poacher. Often obscured among the dense algae, this thin creature may swim near pilings or floats from where naturalists may identify it in part by the jerky sculling movement produced by its pectoral fins. Tiny, planktonic animals comprise most of the slow-moving tubenose poacher's diet while potential predators often ignore its hard, armour encased body.

Do try a tubenose poacher; you may well be the first to do so.

145 Smooth Alligatorfish

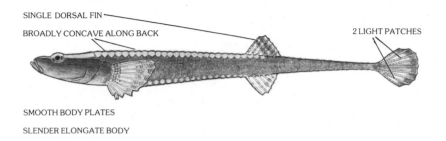

SINGLE DORSAL FIN

BROADLY CONCAVE ALONG BACK

2 LIGHT PATCHES

SMOOTH BODY PLATES

SLENDER ELONGATE BODY

SPECIES: *Anoplagonus inermis* —from the Greek *an*, *plac* and *agonus*, meaning "without," "scales," and "without joints;" as well as the Latin *inermis*, "unarmed."

ALTERNATE NAMES: smooth poacher, smooth sea-poacher.

MAXIMUM RECORDED SIZE: 15 cm (6 inches).

DISTRIBUTION: Point Arena, central California, to the Gulf of Alaska. Korean records exist, but may be for another species.

It is very doubtful that any angler ever has or ever will catch a smooth alligatorfish—a thin, bony, elongate species, unsuitable for bait.

An observant diver may very easily swim past a smooth alligatorfish without seeing it at all or mistaking it for a tiny, submerged stick lying on the sea floor. Usually this dark brown creature rests motionless upon sandy bottoms near large, rocky outcroppings. Look for it also on bottoms littered with woodchips from log booms. Living as shallow as 5 m (16 feet), the slender smooth alligatorfish may occasionally have growing upon its stiff body a coating of unicellular brown plants called diatoms.

When sorting through their valuable catches, shrimp trawlers sometimes find a large smooth alligatorfish crammed into their nets just hauled up from a flat boulder-strewn bottom. This inconspicuous poacher may also occasionally venture into prawn traps, attracted either by the bait or by the tiny creatures which have preceded them. Of no direct value to the commercial fisherman, the smooth alligatorfish undoubtedly falls prey to larger, economically important groundfish.

Because the smooth alligatorfish has been taken in water as shallow as 5 m (16 feet), a keen-eyed wharf bound naturalist might just conceivably see a specimen, though none have as yet been recorded as living in tidepools.

Too small and hard for serious dining.

146 Kelp Poacher

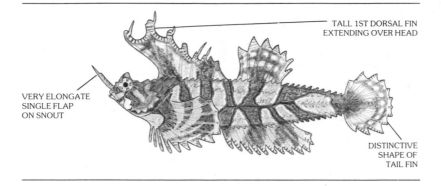

TALL 1ST DORSAL FIN
EXTENDING OVER HEAD

VERY ELONGATE
SINGLE FLAP
ON SNOUT

DISTINCTIVE
SHAPE OF
TAIL FIN

SPECIES: *Agonomalus mozinoi*—from the Greek *agonus* and *omalus*, meaning "without joints" and "even" and honouring Jose Mozino, an early North American naturalist.

ALTERNATE NAMES: none.

MAXIMUM RECORDED SIZE: 8.9 cm (3.5 inches).

DISTRIBUTION: Monterey, central California, to Dixon Entrance, northern British Columbia.

Because the tiny kelp poacher has a tiny mouth, it is extremely unlikely that an angler would catch it.

Amazingly, researchers collected the elusive kelp poacher for the first time in the 1970s, illustrating that significant natural history discoveries still await mankind—particularly divers. With its large, long dorsal fin extending out over its head, exaggerated snout spine, and floppy pectoral fins which enable it to "crawl," this dragon-like creature is surely the most difficult of all Pacific Northwest fishes to see amid its surf-swept, surging open coast habitat and among the dense growths of colourful surrounding seaweeds. Divers in California found this beautifully ornate fish to be most active nocturnally, an observation apparently not yet verified for northern specimens.

Commercial fishermen never harvest the shallow surf-swept habitat of the tiny, unsaleable kelp poacher.

Keen-eyed beachcombers found the strange-looking but well-camouflaged kelp poacher in open coast tidepools and transported these first few specimens to the Vancouver Public Aquarium. Aquarium researcher Dr. Jeff Marliave watched closely the next December as a pair courted and spawned, producing small clusters of between 6 and 25 adhesive, spherical, bright red fertilized eggs. Deposited in an empty barnacle casing and amid the coils of a calcareous tube worm, these fertile eggs hatched into tiny, but distinctive larvae.

Who would eat a tiny and armour-plated "sea dragon"?

182

147 Rockhead

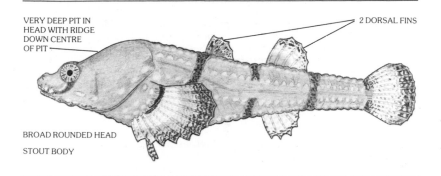

VERY DEEP PIT IN HEAD WITH RIDGE DOWN CENTRE OF PIT

2 DORSAL FINS

BROAD ROUNDED HEAD

STOUT BODY

SPECIES: *Bothragonus swani*—from the Greek *bothros* and *agonus*, meaning "pit" and "without joints;" and honouring early naturalist James G. Swan, of Port Orchard, Washington.

ALTERNATE NAMES: deep-pitted poacher, deep-pitted sea-poacher, pitted poacher, pithead poacher.

MAXIMUM RECORDED SIZE: 8.9 cm (3.5 inches).

DISTRIBUTION: Diablo Cove, Sonoma County, central California, to Kodiak Island, Gulf of Alaska.

The rockhead is really of no significance to the average angler, primarily because of its small stature and tiny mouth, as well as a stiff, hard body that renders it valueless as bait.

Adventurous, keen-eyed divers exploring along rocky shores directly exposed to Pacific surge may catch a glimpse of the rockhead. Look carefully, though—it nestles among the cracks and crevices at depths less than 20 m (66 feet), but it is well camouflaged in the colourful plant and animal life of its surroundings.

Because a rocky, shallow habitat protects it from all commercial enterprises and equipment, the rockhead is never even an incidental catch. Besides, it is of no direct economic value at present.

Rocky or pebble-bottomed tidepools of the exposed coast provide an intertidal habitat for this poacher. Slow-moving and often stationary, when once observed, it is easily captured by hand or dipnet. Small intertidal shrimps and shrimp-like creatures usually become the primary diet of this fish. Pick a rockhead from out of the water and listen closely; it may produce low-pitched vibrations or humming sounds.

Of no culinary value whatsoever.

THE SNAILFISHES
(Family: Liparidae)

Of the one hundred and fifteen known living fishes comprising the Liparidae, including the lumpsuckers, lumpfishes and snailfishes, or sea snails, at least twenty-five species flourish in the Pacific Northwest. Members of this entirely marine family colonize from the tidepools of the splash zone to the abyssal haunts in the great oceanic trenches. While a few species, without sucking discs, swim in midwater well above the sea floor, bottom habitats are home for most liparids. Temperate and frigid seas of the North Pacific provide shelter for nearly three-quarters of the world's snailfishes; the remainder live in the North Atlantic, Arctic and Antarctic Oceans, as well as in tropical waters which are very deep and cold.

Despite centuries of collecting, no fossil liparids have yet been unearthed so science still can only speculate about their origins and study their closest relatives, the rockfishes, greenlings, sculpins, sablefishes and poachers—the mailcheeked fishes. Primarily small fishes, seldom growing longer than 30 cm (12 inches), the plump-looking snailfishes interest man very little economically. The North Atlantic lumpfish, however, produces excellent caviar and, incidentally, happens to be the world's most massive liparid at 60 cm (24 inches).

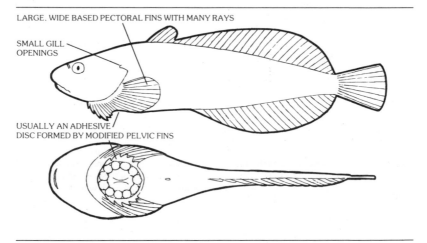

LARGE. WIDE BASED PECTORAL FINS WITH MANY RAYS

SMALL GILL OPENINGS

USUALLY AN ADHESIVE DISC FORMED BY MODIFIED PELVIC FINS

Deep water Pacific Northwest snailfishes not included here in the species accounts are the smalldisc snailfish, with its large eyes; the blacktail snailfish, with its black tail; the abyssal snailfish, with its high dorsal fin; the tadpole snailfish, with no sucking disc and its anus located directly below its eyes; and the prickly snailfish, with no sucking disc and its anus located directly below its gills. Others not included here are the smooth lumpsucker, a northern bulbous form with a single dorsal fin; the ribbon snailfish, with its dorsal fin barely reaching the tailfin; the slimy snailfish, with, at the front of its dorsal fin, a smoothly rounded lobe that is always visible or does not fold; and the ringtail snailfish, with a white band at the base of the tail and a dorsal fin with an undulating outline.

Several recent additions to the snailfish fauna of the Pacific Northwest include the deep water species: the spiny snailfish; the smallfin snailfish; the pygmy snailfish; the bigtail snailfish; the bigpored snailfish; the toothless snailfish; the swellhead snailfish; the pink snailfish; and the slim snailfish.

148 Pacific Spiny Lumpsucker

2 DORSAL FINS

STOCKY BODY AND HEAD
COVERED WITH SPINY CONE-SHAPE PROTUBERANCES

VENTRAL SUCKING DISC
FRINGED WITH THICK MARGIN

SPECIES: *Eumicrotremus orbis*—from the Greek *eu*, *micro* and *tremus*, meaning "truly," "small" and "aperture;" and the Latin *orbis*, "sphere."

ALTERNATE NAMES: spiny lumpsucker.

MAXIMUM RECORDED SIZE: 12.7 cm (5 inches).

DISTRIBUTION: Puget Sound, northern Washington, to St. Lawrence Island, throughout the Bering Sea, along the Aleutian chain to Sakhalin and the Kurile Islands, USSR.

The active Pacific spiny lumpsucker, itself probably never caught on hook and line, would perhaps function as bait—if the cold-hearted angler could actually impale such a cute little fish upon the tackle.

A chance of sighting the lovable Pacific spiny lumpsucker makes worthwhile a shallow summer or autumn dive into a bay choked with eelgrass or kelp. Watch for this rotund little creature resting with its sucking disc attached to something solid such as a plant, a sunken log, or a piece of shell. Dislodge the fish and notice it swim about like an underwater helicopter or miniature ping-pong ball with fins. A slow swimmer, easily tracked by the diver, this stubby creature always appears to be gasping even when at rest. The larger pale-green female has more massive and numerous shield-like tubercles than her yellow or mauve mate, and in late winter lays as many as 200 adhesive, cherry-red eggs. Tiny, recognizable young are often plentiful in eelgrass communities during April and May.

Occasionally a commercial shrimper trawling depths less than 150 m (495 feet) may unintentionally drag up the unmarketable Pacific spiny lumpsucker.

Seldom seen unless it moves, the Pacific spiny lumpsucker commonly swims among the eelgrass and kelp or rests attached to pilings or rocks. It may approach lights at night when it is easily dipnetted. This fish is amusing because its almost invisible fins propel its globular body as it huffs and puffs about in no apparently predictable direction.

Could anyone actually eat the unbelievable Pacific spiny lumpsucker? Never!

149 Tidepool Snailfish

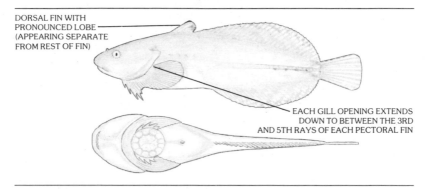

DORSAL FIN WITH PRONOUNCED LOBE (APPEARING SEPARATE FROM REST OF FIN)

EACH GILL OPENING EXTENDS DOWN TO BETWEEN THE 3RD AND 5TH RAYS OF EACH PECTORAL FIN

SPECIES: *Liparis florae*—from the Greek *liparos*, meaning "sleek skinned;" and honouring Flora H. Green of Stanford University.

ALTERNATE NAMES: shore liparid.

MAXIMUM RECORDED SIZE: 12.7 cm (5 inches).

DISTRIBUTION: Point Conception, southern California, to the Bering Sea coast of Alaska.

The tidepool snailfish is unlikely to seize even the tiniest baited hook. It would be a poor live bait for bottom-fishing because it would no doubt immediately attach itself to the nearest object.

Snorkelers might find the often olive-green tidepool snailfish because it lives in very shallow water, usually nestling among the colourful marine plants growing there. However, because this small, shy fish hides extremely well, the diver must search most diligently for it. Overturning submerged rocks often briefly exposes specimens, but they quickly dart off and wriggle beneath nearby shelter. The well-camouflaged tidepool snailfish, undisturbed, rests attached by its suction disc and with its tail curled along its body like a reclining puppy.

Little commercial harvesting occurs intertidally or in shallow rocky habitats where the unsaleable tidepool snailfish lives; consequently, it never shows even as an incidental catch.

Very careful observation among kelps and other marine plants in tidepools or along the sea's edge may sometimes produce a glimpse of the inconspicuous tidepool snailfish. The best way to view and study a specimen closely, however, is either by dipnetting it from among the algae, or by overturning rocks and exposing a specimen as it clings to the underside of a stone. Foraging primarily upon small shrimps and their kin, the pudgy tidepool snailfish can readily change its olive colour either darker or lighter.

The slimy appearance of even the largest tidepool snailfish discourages seafood fanciers.

150 Spotted Snailfish

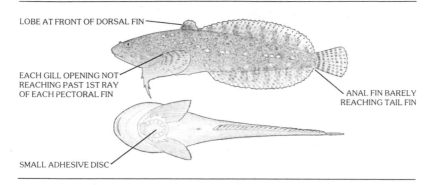

LOBE AT FRONT OF DORSAL FIN

EACH GILL OPENING NOT
REACHING PAST 1ST RAY
OF EACH PECTORAL FIN

ANAL FIN BARELY
REACHING TAIL FIN

SMALL ADHESIVE DISC

SPECIES: *Liparis callyodon* —from the Greek *liparos*, *callos* and *odons*, meaning "sleek skinned," "beautiful" and "tooth."

ALTERNATE NAMES: Pallas's liparid.

MAXIMUM RECORDED SIZE: 12.7 cm (5 inches).

DISTRIBUTION: Oregon to the western Aleutian Islands and the Bering Sea coast of Alaska.

 The tiny-mouthed spotted snailfish lives in very shallow intertidal habitats, much of the time hiding under rocks where it would seldom encounter baited hooks. Its dark colour and shelter-seeking behaviour makes it a poor bottom-fishing bait, alive or dead.

 Alert night divers cruising along shallow, sandy-bottomed beaches with profusely growing eelgrass or scattered rock might catch a brief glimpse of the small spotted snailfish as it actively forages for food. By day, however, the wary creature usually hides beneath any available shelter. If uncovered, the elusive spotted snailfish generally rockets off to wriggle beneath nearby plants, rocks or sunken logs.

 Only commercial fishermen working from shore would ever encounter the unmarketable spotted snailfish.

 Strolling along boulder-strewn beaches at low tide, a curious beachcomber might locate the greenish-gray or brown spotted snailfish by overturning rocks, shells or other shelter. To break the suction of its ventral sucking disc and dislodge this tenacious creature, pull it steadily, firmly, but gently along the surface of its shelter. On a pleasant night, take a bright light and search for the spotted snailfish's dark, tadpole-like form swimming slowly and gracefully as it weaves its way through thick mats of emerald eelgrass. When eventually coming to a halt, this pudgy creature almost always sucks onto any solid object, curls its tail around and along the side of its body and finally lowers its fins.

 Too small and slimy-appearing.

187

151 Slipskin Snailfish

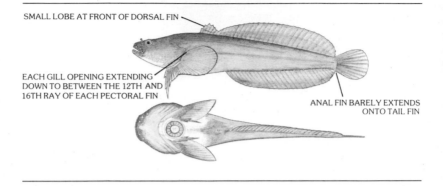

SMALL LOBE AT FRONT OF DORSAL FIN

EACH GILL OPENING EXTENDING DOWN TO BETWEEN THE 12TH AND 16TH RAY OF EACH PECTORAL FIN

ANAL FIN BARELY EXTENDS ONTO TAIL FIN

SPECIES: *Liparis fucensis*—from the Greek *liparos*, meaning "sleek skinned;" and emphasizing the Strait of Juan de Fuca.

ALTERNATE NAMES: Juan de Fuca liparid.

MAXIMUM RECORDED SIZE: 18 cm (7 inches).

DISTRIBUTION: Point San Simeon, northern California, to southeastern Alaska.

A small creature with a tiny mouth, the inconspicuous slipskin snailfish very seldom dangles from the hook of an angler's line either as a catch or as bait.

On that next dive, turn over rocks and sunken logs or look inside bottles and shells to perhaps reveal a shy, secretive slipskin snailfish. Typically this variably coloured snailfish spends most of its time hiding and, when exposed by a curious aquanaut, will invariably rush to other nearby shelter. During its spawning season in early spring to late summer, a large gravid female slipskin snailfish may lay as many as 4,800 tiny eggs forming doughnut-shaped clusters. After accompanying the prolific female and fertilizing these pink to orange eggs, the attentive male remains to guard them, possibly in an empty mussel shell. Upon hatching, the tiny larvae are transparent and bear little resemblance to the adult.

While of no direct economic value, the soft slipskin snailfish nevertheless may come into the realm of some commercial fishermen anyway. Often seeking small shrimp-like prey, this graceful creature steals into prawn traps at depths from 6 to 388 m (20 to 1,280 feet). Sometimes shrimp trawls haul it to the surface too.

The usually deep-dwelling slipskin snailfish does not live around wharves, jetties, and pilings, nor in tidepools where surface-bound naturalists might see it.

Too small and soft for serious dining.

152 Marbled Snailfish

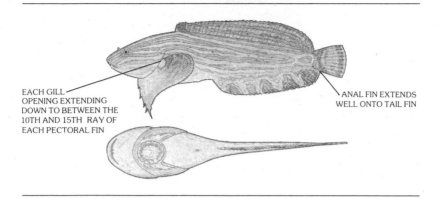

EACH GILL OPENING EXTENDING DOWN TO BETWEEN THE 10TH AND 15TH RAY OF EACH PECTORAL FIN

ANAL FIN EXTENDS WELL ONTO TAIL FIN

SPECIES: *Liparis dennyi*—from the Greek *liparos*, meaning "sleek skinned;" and honouring the late Charles L. Denny of Seattle, Washington.

ALTERNATE NAMES: Denny's liparid.

MAXIMUM RECORDED SIZE: 30 cm (12 inches).

DISTRIBUTION: Puget Sound, northern Washington, to Unimak Island, in the Aleutian chain, Alaska.

Because the tiny-mouthed marbled snailfish usually lives at considerable depths and on soft substrates, it probably never encounters baited hooks lowered by sport anglers. Its often sedentary habitats and well-camouflaged body combine to make it a poor bait.

As yet SCUBA divers rarely find the variably marked marbled snailfish at accessible depths and much remains to be discovered about the ecology and habits of this seldom-seen fish.

Although the small marbled snailfish is of no direct economic value, it often wriggles about among the marine life brought aboard vessels that trawl or trap shrimps. This large snailfish apparently prefers sandy or silty bottoms at depths between 10 and 225 m (33 and 743 feet) where fine-meshed nets easily dislodge it from scattered rubble onto which it has adhered. In captivity, healthy marbled snailfish either suck onto solid objects while resting or swim slowly and sinuously about very close to the bottom, where they often appear to nearly bump into shells, rocks or sunken logs but stop just in time. The mature female swells with tiny pink eggs during the winter, presumably for an early spring spawning.

Beachcombers or dockside observers are unlikely to observe the secretive marbled snailfish because it flourishes well below the surface.

The soft, fine-grained flesh of even the largest marbled snailfish excites few gourmets.

153 Lobefin Snailfish

DISTINCT LOBE AT FRONT
DORSAL FIN

TINY EYES

EACH GILL OPENING
EXTENDING ONLY
AS FAR DOWN
AS THE 4TH
RAY OF EACH
PECTORAL FIN

SPECIES: *Polypera greeni*—from the Greek *poly* and *pera*, meaning "many" and "pouches;" and honouring early naturalist Ashdown Green.

ALTERNATE NAMES: Green's liparid.

MAXIMUM RECORDED SIZE: 31 cm (12.2 inches).

DISTRIBUTION: Washington, to the Bering Sea coast of Alaska.

A wharf-bound angler using a tiny hook baited with small, live, shrimp-like creatures lowered into the dangling kelp might catch a lobefin snailfish. But the chances are slim, and why bother anyway?

Snorkelers as well as SCUBA enthusiasts have access to the large lobefin snailfish because it commonly cowers among the dense summer growths of kelp trailing from floats and pilings. Even a sharp-eyed diver must look for it most intently, however, because the golden brown colour of this scaleless fish matches identically that of the algae onto which it adheres, with its tail curled around and along the side of its body. During a diligent search, move kelp fronds carefully; if startled, the shy lobefin snailfish will dart away.

Occasionally a commercial fisherman might find the unmarketable lobefin snailfish attached to a net float or other parts of commercial gear fished near the surface.

Although the slippery lobefin snailfish—with the front part of its dorsal fin appearing to be almost a separate lobe—lives commonly among the marine plants which grow from floats or pilings, it is very difficult to see. A careful scoop with a fine-meshed dipnet may capture an unsuspecting specimen. First lower the net down until the meshed hoop circles up under the end of a large kelp frond onto which a fish might perhaps adhere. Don't touch the fronds before lifting the net or the fish might escape, then, when ready, quickly pull up the net and hope for a snailfish.

Too soft and fine-grained of flesh for serious dining.

154 Showy Snailfish

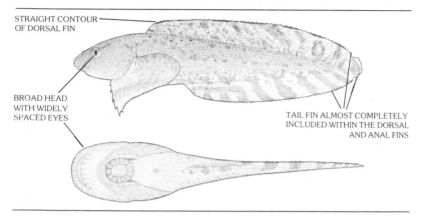

STRAIGHT CONTOUR
OF DORSAL FIN

BROAD HEAD
WITH WIDELY
SPACED EYES

TAIL FIN ALMOST COMPLETELY
INCLUDED WITHIN THE DORSAL
AND ANAL FINS

SPECIES: *Liparis pulchellus*—from the Greek *liparos*, meaning "sleek skinned" and the Latin *pulchellus*, "pretty."

ALTERNATE NAMES: shorttail snailfish, continuous-finned liparid.

MAXIMUM RECORDED SIZE: 25 cm (10 inches).

DISTRIBUTION: Monterey Bay, central California, to the Bering Sea, along the Aleutian chain to Peter the Great Bay, USSR.

The tiny-mouthed showy snailfish lives upon soft substrates in relatively deep water where it probably never encounters bait offered by anglers.

Periodically a secretive showy snailfish will venture out from under solid shelter and cruise slowly over sandy bottoms at depths below 10 m (33 feet); this occurs more often nocturnally and is therefore usually observed by night divers. By day this attractive fish adheres to the undersides of rocks or huddles inside empty bottles, cans and other discarded man-made paraphernalia. Late winter or early spring finds large, gravid, female showy snailfish laying up to 9,200 eggs on the sea floor. After hatching, the tiny transparent larvae rise to the surface and swim feebly about in search of minute shrimp-like prey. Later, as juveniles, then much later as adults, these sinuously swimming snailfish feed heavily upon tiny crabs, shrimps, marine worms, small flounders, and even the young of their own kind. For a showy snailfish, five years is a ripe old age, attained only if predators such as the various flounders don't devour it.

Fine-meshed nets towed by shrimp trawlers frequently dislodge, then drag the striped or spotted showy snailfish to the surface from depths as great as 183 m (600 feet). Commercial fishermen sort it from the valued shrimps and then discard it.

Its pattern of stripes and spots and its slow, easy, gentle cruising make this a beautiful fish to watch—in a public aquarium.

A chef would be hard pressed to prepare a meal from the small showy snailfish.

155 What Species Is It

Several years ago this photograph was taken during a memorable night dive at Copper Cove, near Vancouver—and mystery still shrouds the identity of the creature. Unfortunately, the photographer was unable to collect that specimen and another has not been seen, in spite of the diver returning many times to this popular night dive site. An exciting event that might have easily occurred during anyone's dive, it emphasizes the unexpected or unexplained in the study of Pacific Northwest marine fishes. We suspect this silvery fish to be one of the deep water snailfishes, perhaps a tadpole snailfish, *Nectoliparis pelagicus*, which strayed into unusually shallow water.

156 Undescribed Snailfish

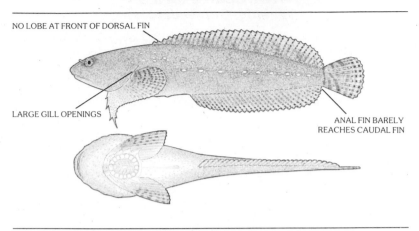

NO LOBE AT FRONT OF DORSAL FIN

LARGE GILL OPENINGS

ANAL FIN BARELY
REACHES CAUDAL FIN

Including this as yet undocumented and un-named species of snailfish in this volume illustrates that the knowledge of Pacific Northwest marine fishes is still far from complete. While aware of this "new" fish, scientists require numerous undamaged specimens in order to describe and catalogue it properly. The few specimens now present in university collections were obtained in the shallow or intertidal waters off Stanley Park, Vancouver. A length of about 10 cm (4 inches) is, so far, a maximum for this little-known snailfish.

THE CLINGFISHES
(Family: Gobiesocidae)

The family of clingfishes, scientifically designated the Gobiesocidae, is a moderately large one of nearly 100 known living species, most of which live intertidally or in shallow, rocky coastal regions. There are very few deep water marine species, and only a few fresh water types live in swift-flowing and hard-bottomed streams. The cover-seeking gobiesocids inhabit primarily tropical and temperate waters and two are known from along Pacific Northwest shores, including the flathead clingfish, one of the world's larger species.

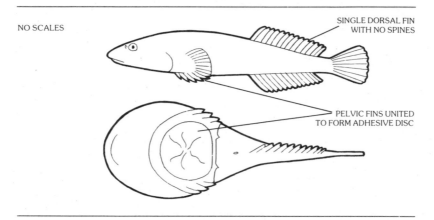

NO SCALES

SINGLE DORSAL FIN
WITH NO SPINES

PELVIC FINS UNITED
TO FORM ADHESIVE DISC

The conspicuous sucking disc, formed by the modification of the pelvic fins and located under the belly area of each species, is the most significant gobiesocid trait. By adhering to solid objects with such suction organs, these resourceful creatures maintain their positions in swift currents or against crashing waves. The organs also allow these animals to conserve energy while in active, turbulent habitats inhospitable to many other fishes. Although a superficial comparison of clingfish sucking discs to the discs of the snailfishes immediately suggests a close relationship between the two groups, closer anatomical study reveals many significant differences between these fishes. A biological theory called "convergent evolution" states that two different creatures or groups of creatures can independently evolve structures which are similar and which fulfill identical functions. This theory, therefore, indicates that despite similar sucking discs, snailfishes and clingfishes are not closely related.

Paleontological studies and dating indicate that gobiesocid ancestors clung tenaciously to rocks in shallow seas at least 12 million years ago, in Miocene times. At present, though, the inconspicuous clingfishes have no truly close relatives.

157 Kelp Clingfish

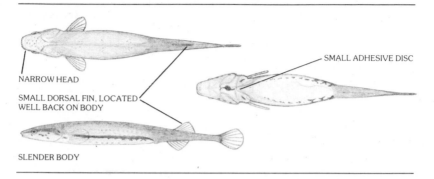

NARROW HEAD

SMALL DORSAL FIN, LOCATED WELL BACK ON BODY

SMALL ADHESIVE DISC

SLENDER BODY

SPECIES: *Rimicola muscarum*—from the Latin *rima*, *cola*, and *muscarum*, meaning "crevice," "inhabiting," and "fly."

ALTERNATE NAMES: slender cling-fish.

MAXIMUM RECORDED SIZE: 7 cm (2.8 inches).

DISTRIBUTION: Bahia Todos Santos, southern Baja California, Mexico, to Goose Island, on the central coast of British Columbia.

Too small to be caught with normal fishing tackle, the minute kelp clingfish is a poor bait fish because of its camouflage coloration and its habit of sucking onto solid objects.

While very common in shallow, weedy bays adjacent to direct Pacific surge, the emerald green and yellow or golden brown kelp clingfish proves to be an elusive quarry for the diver because it adheres to marine plants which are the same colours as itself. Look closely at individual eelgrass plants or kelp fronds for this inconspicuous creature's slender, motionless form. Good luck!

Commercial fishermen do not operate in the surf-swept habitat of the unmarketable kelp clingfish. The only people interested in this tiny species are a few scientists or eager aquarists.

Although rarely frequenting tidepools, the sometimes speckled kelp clingfish flourishes in exposed eelgrass and kelp-choked shallows in which curious beachcombers may wade. The plant life dangling from wharves and pilings erected in these locales may harbour the kelp clingfish. Having dipnetted a specimen, place the captive in a glass container for a closer look. When the tiny fish attaches itself to the transparent surface, easily see the throb of its red minute heart. In summer, if your captive is a gravid female, her eggs will be readily visible inside her body as tiny green spheres. An adult male will have an elongated genital papilla and a groove from his sucking disc to his anal fin.

Too small for serious dining.

158 Northern Clingfish

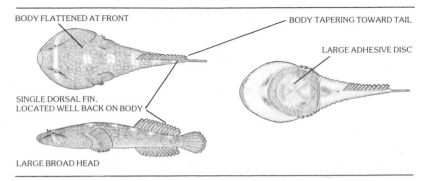

BODY FLATTENED AT FRONT

BODY TAPERING TOWARD TAIL

LARGE ADHESIVE DISC

SINGLE DORSAL FIN,
LOCATED WELL BACK ON BODY

LARGE BROAD HEAD

SPECIES: *Gobiesox maeandricus*—from *goby* and *esox*, the popular ancient names for two fishes, the goby and the pike; and the Latin *maeandricus*, "meandering."

ALTERNATE NAMES: flathead clingfish, common cling-fish.

MAXIMUM RECORDED SIZE: 16 cm (6.5 inches).

DISTRIBUTION: Guadalupe Island and northern Baja California, Mexico, to Revillagigedo Island, southeastern Alaska.

Because the darkly-coloured northern clingfish hides under cover virtually its entire life, anglers very seldom catch it. Because of its tendency to hide under rocks and because of its colour, it makes poor live bait for bottomfishermen.

Very infrequently sighted as it scurries from under one object to another, the northern clingfish attaches itself to the undersides of rocks or sunken logs. On turning over small objects to find a specimen, a frustrated diver may find it will move to the other side of the shelter in a "hide and seek" fashion. Solve this problem by picking up the sheltering object, if it is small enough, and isolating the northern clingfish.

Of no direct economic value, the northern clingfish's small size, secretive habits and shallow or intertidal habitat make it a rare incidental catch.

Beachcombers who overturn rocks or boulders left exposed at low tide commonly encounter the broad-headed northern clingfish clinging tenaciously to its shelter. Dislodge the fish by sliding its body along the shelter at the same time you're pulling it free from the rock or boulder. Various small snails, worms, and shrimp-like creatures sharing the rocky abode of this fish also form much of its diet. During spring a ripe female deposits adhesive yellow eggs to the undersides of intertidal rocks where the embryos develop and hatch within a few weeks. The male guards these eggs and may fertilize more from successive females at the same site.

Too small for serious dining.

THE LEFTEYE FLOUNDERS
(Family: Bothidae)

Containing about 200 recognizable living species of fishes, including the small sanddabs of the Pacific Northwest, the Bothidae is a very large family of bottom dwelling flatfishes popularly known as lefteye flounders. Members of this strictly marine group live in all seas from tropical to boreal latitudes. By attaining a maximum length of 152 cm (5 feet), the species known as the California halibut ranks as the family giant: most bothids only grow to less than half that size.

Curiously, a newborn lefteye flounder begins life near the surface with a symmetrical body and an eye on each side of its head, not unlike that of any other fish. After a definite species-specific period though, two very dramatic changes begin: the migration of the right eye to the left side of the head, beside the other eye; and a corresponding and necessary twisting of the skull. During this dramatic transition, the strange-looking young bothid swims with an ever-increasing sideways tilt to compensate for the ongoing anatomical changes. By the time this eye migration process is completed, the body of the young fish has also deepened noticeably and becomes darkly pigmented, but almost always only on its left or "eyed" side. Now resembling its parents, the young lefteye flounder takes up permanent residence on the bottom, either lying flat upon its right or "sightless" side or swimming about near the bottom with its pigmented side up and propelled by undulating body movements.

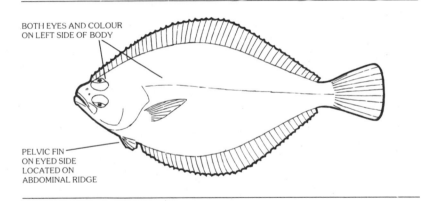

BOTH EYES AND COLOUR
ON LEFT SIDE OF BODY

PELVIC FIN
ON EYED SIDE
LOCATED ON
ABDOMINAL RIDGE

Other flatfish groups, such as the Pacific Northwest's own righteye flounders, are closely related to the bothids, and ichthyologists believe they all originally descended from an ancient, symmetrical perch-like ancestor that began lying on its side. Commercial fishermen the world over profitably harvest lefteye flounders.

Although apparently reported from as far north as British Columbia, the California halibut, with its large mouth and jaws that extend well back behind the eyes, is not abundant or commonly harvested in the Pacific Northwest.

159 Speckled Sanddab

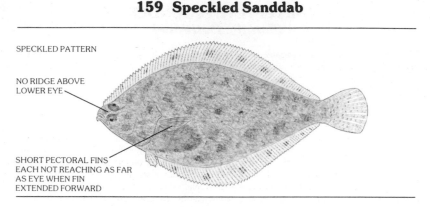

SPECKLED PATTERN

NO RIDGE ABOVE
LOWER EYE

SHORT PECTORAL FINS
EACH NOT REACHING AS FAR
AS EYE WHEN FIN
EXTENDED FORWARD

SPECIES: *Citharichthys stigmaeus*—from the Greek *citharos*, *ichthys* and *stigmaeus*, meaning "rib," "fish" and "speckled."

ALTERNATE NAMES: speckled sand dab, Catalina sand dab.

MAXIMUM RECORDED SIZE: 17 cm (6.8 inches).

DISTRIBUTION: Bahia Magdalena, southern Baja California, Mexico, to Montague Island, Gulf of Alaska.

The small and common speckled sanddab readily seizes such baits as tiny shrimps, marine worms or little pieces of fish. Considering it an irritating nuisance, shore or pier-bound fishermen particularly encounter this active flatfish resting on sandy and muddy bottoms. It can be useful as bait.

The ultimate in camouflage, a resting speckled sanddab is very difficult to notice until it moves. And even then, this shallow water denizen may be difficult to relocate once it stops again. Frustrated underwater photographers may take technically perfect pictures but often view their speckled sanddab "shots" with considerably dismay: a lack of contrast between the subject and its sandy background gives images which look like mere gray and lifeless sand. Look for the deceptive speckled sanddab patiently—especially at depths less than 15 m (50 feet).

Although found at depths to 549 m (1,800 feet), the thin speckled sanddab is primarily an incidental, unmarketable catch of trawlers commonly dragging their nets much shallower in search of other harvests.

Particularly vulnerable to winged predators such as cormorants, gulls or herons, the gray speckled sanddab readily flees before beachcombers who wade through shallows on tidal flats. Tolerant of temperatures less than 13°C and spawning from March to September, this fish apparently does not usually survive beyond 4 years.

While tasty, the speckled sanddab's flesh contains many small bones and very little is available on each specimen.

160 Pacific Sanddab

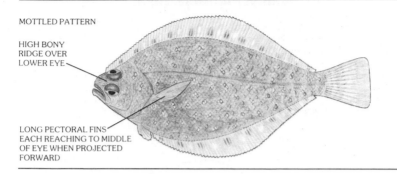

MOTTLED PATTERN

HIGH BONY
RIDGE OVER
LOWER EYE

LONG PECTORAL FINS
EACH REACHING TO MIDDLE
OF EYE WHEN PROJECTED
FORWARD

SPECIES: *Citharichthys sordidus*—from the Greek *citharos* and *ichthys*, meaning "rib" and "fish;" and the Latin *sordidus*, "dull colour."

ALTERNATE NAMES: mottled sanddab, mottled sand dab, soft flounder, melgrim.

MAXIMUM RECORDED SIZE: 41 cm (16 inches).

DISTRIBUTION: Cabo San Lucas, southern Baja California, Mexico, to the Bering Sea coast of Alaska.

Anglers who bottomfish from boats most often catch the brownish-gray Pacific sanddab because it frequents soft bottoms at depths greater than 15 m (50 feet). Try baits such as marine worms, squid, shrimp, clams or pieces of fish but do not expect a great scrap from this species.

Divers infrequently encounter the Pacific sanddab because it usually thrives in water deeper than that to which most SCUBA aficionados descend. Adult specimens seem to prefer coarser bottoms while the young, which surprisingly inhabit deeper water, usually rest upon silty, fine-grained substrates. Perhaps because the Pacific sanddab may come into shallower water to feed nocturnally, night divers have better chances of viewing this thick-bodied fish.

While trawlers in the Pacific Northwest incidentally take only modest quantities of Pacific sanddab, which is marketed as "sole," this prized species receives much greater attention in California where harvesters may also catch it with hook and line, or hoop nets with baited hooks attached around the rims.

The Pacific sanddab apparently lives at depths inaccessible to surface-bound naturalists and anyone supposedly seeing one is probably seeing the very similar speckled sanddab.

The tasty, flaky flesh of the Pacific sanddab is excellent either baked or fried, especially if cooked fresh. In California, many gourmets consider it a delicacy and some of the finest restaurants there specifically list it on their menus and serve it whole. Chinese immigrants often split a large Pacific sanddab and then dry it for future use.

THE RIGHTEYE FLOUNDERS
(Family: Pleuronectidae)

The righteye flounders, known scientifically as the Pleuronectidae, which means "side family" contains approximately 100 known species, nearly all of which live in the marine waters of the northern hemisphere, in the Pacific and Atlantic Oceans. Although a few species such as the starry flounder and Pacific halibut, may sometimes break the rule, righteye flounder adults have both their eyes on the right sides of their heads. Having two eyes on one side of the head and a broad flat body intimately adapts these fishes for a bottom dwelling existence upon shiftable substrates such as gravel, sand or mud. The Atlantic halibut is the largest family member, reaching 3 m (10 feet) in length and 318 kg (700 pounds).

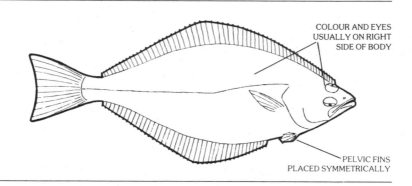

COLOUR AND EYES
USUALLY ON RIGHT
SIDE OF BODY

PELVIC FINS
PLACED SYMMETRICALLY

After closely studying the pleuronectid fossils gathered over many years, paleontologists have determined that ancestral flounders have flourished in the world's seas for at least 50 million years, since lower Eocene times. Other modern day flatfish families, such as the lefteye flounders, are the closest living kin to the righteye flounders. The amazing transformation of young, symmetrical flatfish larvae to asymmetrical adults as described for the bothids on page 197, is essentially the same for pleuronectids—except that the opposite eye, the left one, usually does the migrating.

Inaccessible to most people, several species of righteye flounders live in deep waters of the Pacific Northwest: the roughscale sole, with rows of large rough tubercles on its eyed side; the deepsea sole, with the parts of its body next to the dorsal and anal fins being very thin; the yellowfin sole, a northern species with scaly yellow fins; and the Greenland halibut, with its darkly coloured blind side and big, tooth-filled jaws. Occasionally commercial fishermen catch a naturally occurring hybrid, believed to be the result of inter-breeding between the starry flounder and the English sole. This amazing fish was even given a species name and called the hybrid sole.

161 Pacific Halibut

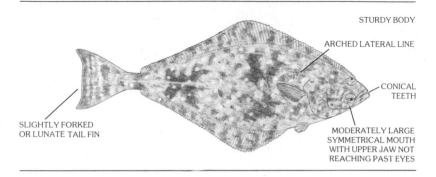

STURDY BODY

ARCHED LATERAL LINE

CONICAL TEETH

SLIGHTLY FORKED OR LUNATE TAIL FIN

MODERATELY LARGE SYMMETRICAL MOUTH WITH UPPER JAW NOT REACHING PAST EYES

SPECIES: *Hippoglossus stenolepis*-from the Greek *hippos*, *glossus*, *steno* and *lepis*, meaning "horse," "tongue," "narrow" and "scale."

ALTERNATE NAMES: halibut.

MAXIMUM RECORDED SIZE: 267 cm (105 inches), and 225 kg (495 pounds).

DISTRIBUTION: Santa Rosa Island, southern California, to St. Lawrence Island and throughout the Bering Sea and south to northern Japan.

The lucky and surprised angler who catches a large Pacific halibut for the first time is often ill prepared and without the necessary heavy tackle. Even salmon fishermen mooching or slowly trolling for chinook sometimes hook this species which has a thick, but streamlined body and lunate, or crescent shaped tail. The large-mouthed Pacific halibut readily seizes baits such as octopus, squid or herring before charging off in long powerful runs. Subdue this large fish *before* hauling it into your small boat because the slapping tail of a large Pacific halibut has maimed and even killed people.

Few divers ever see the distinctly-shaped Pacific halibut because it most often lives in deep and inaccessible water. Any spearfisherman who can take a shot at this impressive beast should consider himself or herself fortunate.

Using set lines, which are bottom-fishing long lines, usually with many octopus-bearing hooks, halibut fishermen harvest carefully controlled amounts along North America's shores. Small Pacific halibut, between 9 and 12 pounds (4.1 and 5.4 kg), sell as "chicks" or "chicken" halibut while very large specimens are sometimes called "whales." In 1932, Canada and the United States founded the International Pacific Halibut Commission to manage this very important fishery.

Shorebound naturalists should visit public aquariums to see the Pacific halibut alive.

Although recent revelations about mercury-contamination have cast a shadow on some stocks, the thick-bodied Pacific halibut remains one of the finest seafoods available in the Pacific Northwest.

162 Arrowtooth Flounder

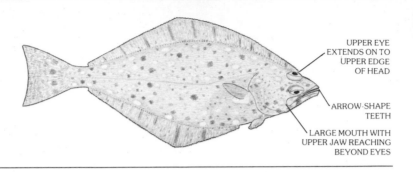

UPPER EYE
EXTENDS ON TO
UPPER EDGE
OF HEAD

ARROW-SHAPE
TEETH

LARGE MOUTH WITH
UPPER JAW REACHING
BEYOND EYES

SPECIES: *Atheresthes stomias*—from the Greek *athor*, *esthio*, and *stoma*, meaning "spike," "eat," and "mouth."

ALTERNATE NAMES: arrowtooth sole, long-jawed flounder, needle-toothed halibut, arrowtooth halibut, American arrowtooth halibut, bastard halibut, french sole, turbot *

MAXIMUM RECORDED SIZE: 84 cm (33 inches).

DISTRIBUTION: San Pedro, central California, to both the Alaskan and Soviet Bering Sea coasts.

Although possessing a large, heavily toothed mouth adapted to seizing small baitfishes such as herring, anchovies, shiner seaperch, walleye pollock and others, the loosely-scaled arrowtooth flounder rarely shows up in anglers' catches.

A species that only rarely invades depths as shallow as 12 m (40 feet), the grayish-brown arrowtooth flounder is most likely to cross the diver's path at night. When on a prawn gathering sortie a diver might see a young and spotted specimen resting very still upon the sandy sea floor. It too may be waiting to ambush tasty prawns, but the hungry arrowtooth flounder also hunts for crabs, squid, or other shrimp.

Although not yet popular with Canadian or American seafood consumers, the large arrowtooth flounder does form a considerable portion of the groundfish catch, particularly in the northern reaches of its distribution. Occasionally, while seeking Pacific halibut at depths to 900 m (2,970 feet), disgusted set liners incidentally haul the unwanted arrowtooth flounder aboard. Instead of using it as mink food or fertilizer, as North Americans sometimes do, Asians eagerly harvest and consume it themselves.

Shallow waters accessible to surface-bound naturalists never harbour the slender arrowtooth flounder.

Even when quickly chilled, the flesh of the arrowtooth flounder turns mushy upon cooking. Perhaps try it in a fish paste along with herbs, cheeses, or other ingredients.

* —incorrect

163 Sand Sole

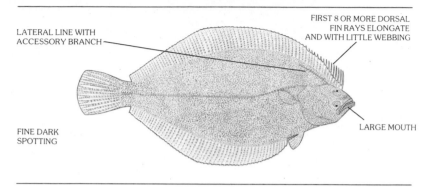

LATERAL LINE WITH
ACCESSORY BRANCH

FIRST 8 OR MORE DORSAL
FIN RAYS ELONGATE
AND WITH LITTLE WEBBING

FINE DARK
SPOTTING

LARGE MOUTH

SPECIES: *Psettichthys melanostictus*—from the Greek *psetta*, *ichthys*, *melas*, and *stichos*, meaning "flounder,""fish," "black," and "specks."

ALTERNATE NAMES: fringe sole, sand flounder, spotted flounder.

MAXIMUM RECORDED SIZE: 63 cm (24.8 inches).

DISTRIBUTION: Point Hueneme, southern California, to the Bering Sea coast of Alaska, near the Alaskan Peninsula.

Try bottomfishing for this moderately large-mouthed species by using its natural food as bait—marine worms, clams, shrimps, herring or anchovies. The gray sand sole commonly rests upon uninterrupted expanses of sand in shallow habitats and is therefore accessible to beach anglers.

With its first few dorsal fin rays being webless this sole commonly lives within easy diving range, but its finely-speckled coloration and burying behaviour make it a very difficult sighting for even the most observant diver. All that a diver often sees is a cloud of silt as a tail beats a retreat over the sea floor and into the murk. In certain geographical locations ugly cancerous tumours cover the bodies of juvenile sand sole but numerous studies have failed to determine an exact cause for such deformations.

Not specifically sought, the finely-marked sand sole forms only a minor portion of the flatfish trawl catch. While known from depths as great as 183 m (604 feet), this distinctive flounder generally prefers much shallower territory and this preference accounts for the minor harvest. Sand sole, marketed with several others as "sole," sells as fillets, either fresh or frozen.

One of the most difficult flounders to see, let alone identify, from an above-water vantage point, the broad and well-camouflaged sand sole rests, usually undetected, along beaches or beneath floats and piers.

Try flounder a l'orange. Place fillets, skin-side down, in a buttered baking pan; add frozen orange juice concentrate, butter, and nutmeg; then bake.

164 Starry Flounder

PROMINENT DARK
BANDS ON DORSAL,
TAIL AND ANAL FINS

ROUGH STAR-SHAPED
PLATES ON BODY

SPECIES: *Platichthys stellatus*—from the Greek *platy* and *ichthys*, meaning "flat" and "fish;" and the Latin *stellatus*, "starry."

ALTERNATE NAMES: grindstone, leatherjacket.

MAXIMUM RECORDED SIZE: 91 cm (36 inches) and 9.1 kg (20 pounds).

DISTRIBUTION: Santa Barbara, southern California, to the Bering Sea and along the Arctic coast to the Coronation Gulf, Northwest Territories; from the Bering Strait to Korea and southern Japan.

A very popular flounder for the baitfisherman angling from wharves, jetties or shorelines, the aggressive starry flounder readily accepts baits such as marine worms, clams, shrimps, mussels and small crabs. Small spinners slowly retrieved near the sandy bottom also attract this good light-tackle battler. Fly fishermen seeking a novel experience can sometimes catch this inhabitant of shallow water when it rises to the surface at night, attracted either by pier lights or the small prey congregating there. The tolerant starry flounder often invades river estuaries and commonly ventures well upriver into fresh water.

Often resting partially or completely buried in soft substrates at depths less than 15 m (50 feet), the grayish starry flounder may also slowly cruise along the bottom in search of prey. Look for it in areas adjacent to eelgrass beds or under wharves.

Working at depths down to 275 m (908 feet), trawlers secure modest numbers of the starry flounder, fillet them and sell them as "sole." Domestic halibut set liners also catch some incidentally. Japanese and Soviet trawl fleets, though, probably take the largest share.

Tiny starry flounders often scatter in all directions before the wading beachcomber who wanders through the shallows of gently sloping tidal flats.

While the rough, leathery skin was once used as emergency sandpaper, it is now discarded and only the tasty flesh retained as an ideal "sole" food.

165 Rock Sole

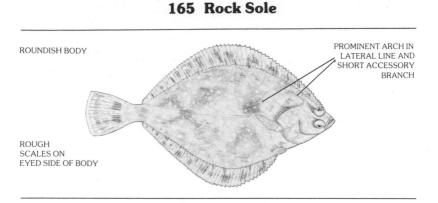

ROUNDISH BODY

PROMINENT ARCH IN
LATERAL LINE AND
SHORT ACCESSORY
BRANCH

ROUGH
SCALES ON
EYED SIDE OF BODY

SPECIES: *Lepidopsetta bilineata*—from the Greek *lepis* and *psetta*, meaning "scale" and "flounder;" and the Latin *bi* and *lineata*, "two" and "lined."

ALTERNATE NAMES: roughscale sole, broadfin sole, broad-fin sole, whitebellied sole, two-lined dab*.

MAXIMUM RECORDED SIZE: 60 cm (24 inches).

DISTRIBUTION: San Nicholas Island, southern California, to the Bering Sea and St. Lawrence Island and then south to Korea and southern Japan.

Very popular with bottomfishing anglers, the brownish rock sole eagerly seizes piling worms, clams and shrimps when fished on sandy, silty or muddy substrates. Artificial lures, too, including trout flies, spinners, and jigs which often foul hook the curious creature, attract this aggressive feeder. A large rock sole can provide good sport on light tackle.

Present throughout all diveable depths, the active rock sole commonly dwells upon sandy or muddy bottoms but may frequently, though briefly, move onto smooth rocky outcroppings. This flounder seldom seems to bury its often rusty-brown body in the sea floor but appears to move about almost constantly in search of prey. Spearfishermen may find a large rock sole a worthy prize.

Highly regarded since the early 1960s, the thick-bodied rock sole is a primary target of the groundfish trawl fleet when fishing its nets down to 366 m (1,200 feet). Land-based processing plants in the Pacific Northwest fillet this species and market it specifically as "rock sole" because of its fine quality.

Young rock sole often flourish around wharves and jetties within easy view of pierside naturalists.

A delicious flavour and firm, flaky texture: a favourite for the ever-popular "fillet of sole"—a dish with many variations.

*—incorrect

166 English Sole

LATERAL LINE WITHOUT HIGH ARCH
AND WITH A LONG ACCESSORY BRANCH

UPPER EYE PARTIALLY
VISIBLE FROM
BLIND SIDE

SOMEWHAT SLENDER BODY
SMALL SCALES

POINTED SNOUT
WITH SMALL
MOUTH

SPECIES: *Parophrys vetulus*—from the Greek *para*, *ophrys* and *vetulus*, meaning "near," "eyebrow" and "old man."

ALTERNATE NAMES: lemon sole, pointed-nosed sole, common sole, California sole.

MAXIMUM RECORDED SIZE: 57 cm (22.5 inches).

DISTRIBUTION: Bahia San Cristobal, central Baja California, Mexico, to Unimak Island in the Aleutian chain, Alaska.

The slippery English sole is common, popular quarry for bottomfishermen who lower baits such as marine worms, clams and small crabs onto sandy or silty substrates. Readily caught from jetties, piers or boats, it provides fair sport for light tackle spincasters and may even seize a slowly-retrieved lure.

One of the most common flounders within the SCUBA diver's domain, the variably-patterned English sole is often difficult to see, especially when completely or partially buried. Often a diver may notice only the "cloud of silt" remaining as a startled specimen darts away. Spearfishermen rarely shoot even a large one.

One of the most heavily exploited groundfish, the valuable English sole is filleted and sold after being hauled from depths as great as 550 m (1,815 feet). Studies indicate that this popular fish exists in numerous distinct, non-intermingling populations along the coast and consequently management of these stocks occurs with reasonable confidence.

Although very difficult to distinguish from the young of several other species, juvenile English sole flourish in shallow bays and tidal flats where they often flee before wading beach strollers. Dockside observers, too, may frequently notice these young resting upon the bottom.

Keep English sole chilled to ensure the best eating of its tasty but fine-grained flesh. Occasionally the flesh of a specimen may be very mushy or milky after cooking because of the spores from parasites. Throw it out.

167 Dover Sole

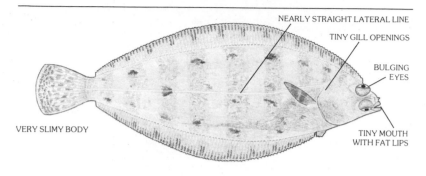

NEARLY STRAIGHT LATERAL LINE

TINY GILL OPENINGS

BULGING EYES

VERY SLIMY BODY

TINY MOUTH WITH FAT LIPS

SPECIES: Microstomus pacificus—from the Greek *micros* and *stomus*, meaning "small" and "mouth;" and emphasizing the Pacific Ocean.

ALTERNATE NAMES: slime sole, slippery sole, shortfinned sole, short finned sole, rubber sole, chinese sole, lemon sole*, smear dab*, tongue sole*.

MAXIMUM RECORDED SIZE: 76 cm (30 inches).

DISTRIBUTION: San Cristobal, central Baja California, Mexico, to the Bering Sea coast of Alaska.

A favourite with many bottomfishermen, the large dover sole may seize baits such as clams, worms or shrimps. Use small hooks for this tiny-mouthed fish and angle in deep water, on sandy or muddy bottoms. A large specimen can provide good light-tackle action.

Often burying itself in the sea floor with only its distinctive, bulbous eyes protruding, the slender dover sole often lives at depths below those scoured by SCUBA divers. Nighttime rather than daytime searches are more rewarding.

Once scorned because of its habit of exuding great quantities of slime which readily spoiled entire catches, the now highly-prized dover sole comprises a large, important share of the retained flatfish trawl catch. After being dragged from soft level bottoms as deep as 1,100 m (3,630 feet), this thick-bodied species is quickly washed, filleted and skinned to eliminate the slime problems. Before the left eye of a dover sole larva completely migrates to the right side of its head and metamorphosis is completed, the young fish may swim at the surface for up to a year.

Not seen near the surface as a recognizable dover sole.

Place dover sole fillets in single layer in buttered baking dish; brush with lemon juice and let stand for ten minutes. Then broil fillets on one side before removing from heat and adding a sauce, parmesan cheese, salad dressing and onion.

* —incorrect

168 Butter Sole

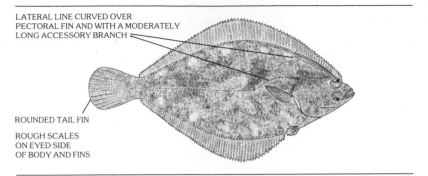

LATERAL LINE CURVED OVER
PECTORAL FIN AND WITH A MODERATELY
LONG ACCESSORY BRANCH

ROUNDED TAIL FIN

ROUGH SCALES
ON EYED SIDE
OF BODY AND FINS

SPECIES: *Isopsetta isolepsis*—from the Greek *isos*, *psetta* and *lepsis*, meaning "equal," "flounder" and "scale."

ALTERNATE NAMES: scalyfin sole, scaly-fin sole, scaly-finned sole flounder, Bellingham sole, Skidegate sole.

MAXIMUM RECORDED SIZE: 54 cm (21.8 inches).

DISTRIBUTION: Ventura, southern California, to the Alaskan Peninsula and Amchitka Island in the Aleutian chain.

Indiscriminate bottomfishing occasionally yields the seldom-caught butter sole, usually from a silty sea floor. It preys particularly upon shrimps, clams, marine worms, sandlance and small herring, so any of these make excellent bait on a smallish hook. Do not expect an exciting battle from this small-mouthed fish, even when it seizes light tackle.

While patrolling over muddy-bottomed areas, an observant diver may sometimes notice the brownish butter sole, but usually only after it has moved. Buried or partially obscured specimens often lie very still and remain undetected by cruising aquanauts. Young, small butter sole haunt shallow, eelgrass-choked bays where often a lack of clear water makes sightings difficult.

Though trawlers do take the sporadically abundant butter sole from all depths to 360 m (1,188 feet), it rates poorly as a marketable commodity and the commercial fisherman often discards it to make room for more valuable species. It occasionally appears as animal food.

Provided the water is clear enough, jettyside strollers or wharf-bound naturalists may occasionally notice this well-camouflaged flounder moving over the bottom, but may have difficulty identifying it as a butter sole.

Over moderate heat, fry butter sole fillets on each side until they are flaky and nicely browned. Remove fish and dry on absorbant paper. In a pan combine garlic salt, potato flakes and some broth. Simmer till thick and then pour over fillets.

169 Flathead Sole

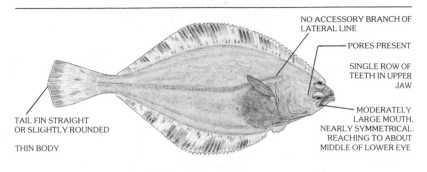

NO ACCESSORY BRANCH OF LATERAL LINE

PORES PRESENT

SINGLE ROW OF TEETH IN UPPER JAW

MODERATELY LARGE MOUTH, NEARLY SYMMETRICAL, REACHING TO ABOUT MIDDLE OF LOWER EYE

TAIL FIN STRAIGHT OR SLIGHTLY ROUNDED

THIN BODY

SPECIES: *Hippoglossoides elassodon* —from the Greek *hippos*, *glossus*, *oides*, *ellas* and *odons*, meaning "horse," "tongue," "like," "small" and "tooth."

ALTERNATE NAMES: paper sole, cigarette paper.

MAXIMUM RECORDED SIZE: 46 cm (18 inches).

DISTRIBUTION: Point Reyes, northern California, and north, throughout the Bering Sea, the Sea of Okhotsk, USSR and the Sea of Japan.

Sometimes while specifically seeking other popular flounders, disappointed baitfishermen catch the sluggish flathead sole over silty or muddy bottoms where it consumes shrimps, clams, marine worms, and small fishes.

While the brownish flathead sole does live upon muddy bottoms at depths as shallow as 5 m (16 feet), it flourishes deeper than most sport divers go. Its habit of burying itself either completely or partially, further hinders observations by aquanauts.

Although the paper-thin flathead sole is often the most abundant flounder hauled up from depths to 550 m (1,815 feet) in trawl nets, it never sells as human food and only rarely as animal food or for reduction. For some as yet unknown reason, a large percentage of specimens in some locales grow nodular or cauliflower-shaped tumors upon their bodies. During the flathead sole's March to May spawning season, a large female might extrude up to 600,000 floating eggs which should hatch about three weeks later. A flathead sole may survive at least 12 years.

The variably-coloured flathead sole seldom invades water shallow enough for the surface-bound enthusiast to view it.

Place a single layer of flathead sole fillets in a baking dish. Marinate in pineapple juice, steak sauce, salt and pepper for 30 minutes. Remove fillets to greased broiling pan and broil for 5 minutes, basting with marinade.

170 Petrale Sole

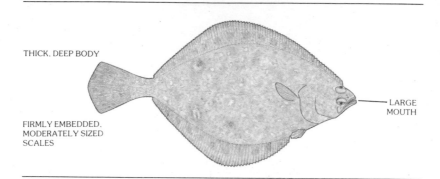

THICK, DEEP BODY

FIRMLY EMBEDDED,
MODERATELY SIZED
SCALES

LARGE
MOUTH

SPECIES: *Eopsetta jordani*—from the Greek *eos* and *psetta*, meaning "morning" and "flounder;" as well as honouring the late David S. Jordan, the "father of American ichthyology."

ALTERNATE NAMES: brill, round-nosed sole, Cape sole, petrale, English sole*.

MAXIMUM RECORDED SIZE: 70 cm (27.5 inches) and 3.6 kg (8 pounds)

DISTRIBUTION: Islas Los Coronados, northern Baja California, Mexico, to the Aleutian chain and the Bering Sea coast of Alaska.

Only those anglers who bottomfish at considerable depths might catch the desirable petrale sole, a large-mouthed flounder that forages heavily upon small fishes such as herring and anchovies. Salmon fishermen who mooch or bottom fish with live bait too close to a soft and level sea floor occasionally take this excellent quarry.

Even the shallow-dwelling juvenile petrale sole usually rests upon substrates well below those commonly patrolled by divers.

Next to the renowned Pacific halibut, the heavy petrale sole has always rated as the number one flounder in the Pacific Northwest. While halibut setliners do incidentally take some petrale sole, the efficient trawlers harvest the bulk of the catch, hauling it from depths to 457 m (1,500 feet). Undoubtedly because of heavy fishing historically, the valuable harvest of this fish has steadily declined in recent years, prompting fisheries biologists in both Canada and the United States to undertake increased studies.

The large petrale sole never enters water shallow enough to permit viewing from the surface.

Just filleting the heavy, thick-bodied petrale sole can be a thrill for the chef anticipating a delicious seafood experience. This flounder is one of the finest species in the Pacific Northwest to eat and even garners specific attention from gourmet gurus like Julia Child who recommends it by name in her famous recipes.

*—incorrect

210

171 Slender Sole

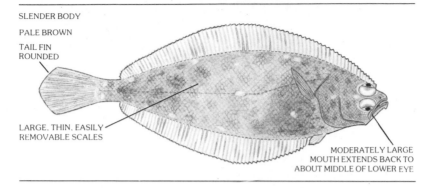

SLENDER BODY

PALE BROWN

TAIL FIN
ROUNDED

LARGE, THIN, EASILY
REMOVABLE SCALES

MODERATELY LARGE
MOUTH EXTENDS BACK TO
ABOUT MIDDLE OF LOWER EYE

SPECIES: *Lyopsetta exilis* —from the Greek *lyo* and *psetta*, meaning "loosen" and "flounder;" and the Latin *exilis*, "slender."

ALTERNATE NAMES: slender flounder, rough sole.

MAXIMUM RECORDED SIZE: 35 cm (13.8 inches).

DISTRIBUTION: Islas Cedros, central Baja California, Mexico, to Alsek Canyon, southeastern Alaska.

The small slender sole lives upon silty or muddy substrates, usually at depths where anglers seldom fish for bottom-dwelling quarry. Small creatures sharing this level and soft sea floor—shrimp, worms and clams—comprise its diet.

Under cover of darkness, the light-brown slender sole may migrate into shoreline habitats, as shallow as 10 m (33 feet), where night-diving prawn seekers may encounter it. Unfortunately for the interested diver, this large-scaled species often digs itself into the sea floor by "swimming on the spot," then letting the silt settle back down onto its body.

Although commercial trawlers, particularly shrimpers with fine mesh nets, take large quantities of slender sole at depths to 500 m (1,650 feet), its small size eliminates it from the seafood market. If not dumped at sea in favour of more valuable cargo, it may reach port as animal food or as matter for reduction. Midwater trawl samples of the abundant slender sole indicate that it sometimes swims well above the bottom in search of floating or planktonic shrimp-like prey. This thin flounder spawns in late winter and by May or June the minute transparent larvae can be very plentiful at the surface.

That a dockside observer would ever see the distinctively thin and narrow slender sole with its easily lost scales is unlikely.

Combine onion, bread crumbs and butter as a stuffing. Mix well and then roll it up in fillets and fasten with toothpicks. Coat the rollups of slender sole with mushroom soup; then wrap them in foil and bake.

211

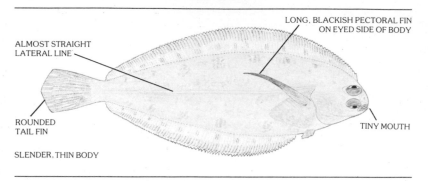

LONG, BLACKISH PECTORAL FIN
ON EYED SIDE OF BODY

ALMOST STRAIGHT
LATERAL LINE

ROUNDED
TAIL FIN

TINY MOUTH

SLENDER, THIN BODY

SPECIES: *Glyptocephalus zachirus*—from the Greek *glyptos*, *cephalos*, *za* and *cheir*, meaning "sculptured," "head," "long" and "hand."

ALTERNATE NAMES: longfin sole, long-finned sole, witch.

MAXIMUM RECORDED SIZE: 59 cm (23.4 inches).

DISTRIBUTION: San Diego, southern California, to the Bering Sea coasts of Alaska and Asia.

Even if the slender rex sole were to nibble at deeply fished baits, its very tiny mouth would be unable to grasp any but the very smallest hook.

Prawn-seeking night divers venturing over gently sloping sandy bottoms sometimes find the pale rex sole being very still on the substrate. If gently coaxed, rather than rudely startled, this distinctive fish will slowly, almost effortlessly, coast along the bottom and erect its characteristic long dark sickle shaped pectoral fin.

Hauled up in abundance from soft substrates at depths as great as 732 m (2,416 feet), the very thin-bodied rex sole comprises much of the total trawl catch of flatfish. However, it is usually discarded in favour of more profitable species and only occasionally reduced for fertilizer. Male rex sole mature by three years of age but the female usually requires an extra two years to ripen for a spawning sometime between January and June. Recent studies show that the amazing larvae of this species may remain near the surface for an entire year while growing to 7.4 cm (3 inches) before descending to their bottom-dwelling existence.

The graceful rex sole lives at depths where non-diving naturalists can never see it.

Do not fillet the thin rex sole—it's too frustrating. Instead, remove its head, fins and viscera, then coat the remainder with flour before frying. Tasty.

173 C-O Sole

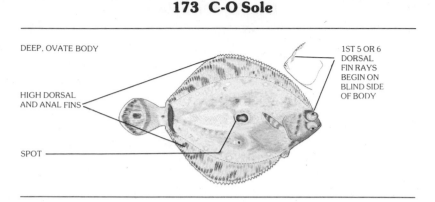

DEEP, OVATE BODY

1ST 5 OR 6 DORSAL FIN RAYS BEGIN ON BLIND SIDE OF BODY

HIGH DORSAL AND ANAL FINS

SPOT

SPECIES: *Pleuronichthys coenosus*—from the Greek *pleuron* and *ichthys*, meaning "side" and "fish;" and the Latin *coenosus*, "muddy."

ALTERNATE NAMES: spot flounder, popeye sole, muddy flounder, mottled turbot, turbot, stinker*.

MAXIMUM RECORDED SIZE: 36 cm (14 inches).

DISTRIBUTION: Bahia San Quintin, northern Baja California, Mexico, to Sitka, southeastern Alaska.

While baitfishing from jetties, piers or the shore, anglers sometimes catch the popeyed C-O sole on small hooks baited with marine worms, clams, and shrimp. A small species, it provides little sport.

So named because of the "initials" on its tail, the brown C-O sole usually lives in shallow water, less than 15 m (50 feet), and often where sandy substrates adjoin rocky outcroppings. Divers tend to see it more than they see some other flounders. It often rests upon or slowly moves among eelgrass, kelps or other marine plants and skillful aquanauts may easily dipnet it or even pick it up by hand. While the variably-marked and coloured C-O sole usually has rich tones of brown on its body, specimens living around the pink coralline algae may have beautiful pink splotches. Consequently, it is one of the few flounders that attracts the discriminating underwater photographer.

While deep-fishing trawl fleets of the Pacific Northwest do not often net the shallow-dwelling C-O sole, their Californian counterparts consider it to be moderately important.

Beachcombers wading through eelgrass beds, or sharp-eyed dockside naturalists peering into the water, occasionally notice the rounded form of the C-O sole effortlessly cruising along the bottom.

Minus its head, fins and entrails, or filleted, the small, round and thick body of the C-O sole is excellent for any recipe calling for "sole."

* —incorrect

174 Curlfin Sole

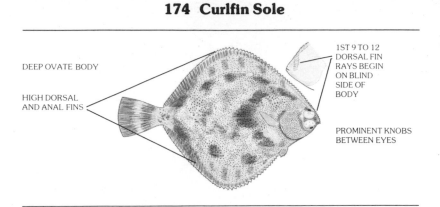

DEEP OVATE BODY

HIGH DORSAL
AND ANAL FINS

1ST 9 TO 12
DORSAL FIN
RAYS BEGIN
ON BLIND
SIDE OF
BODY

PROMINENT KNOBS
BETWEEN EYES

SPECIES: *Pleuronichthys decurrens*—from the Greek *pleuron* and *ichthys*, meaning "side" and "fish;" and the Latin *decurrens*, "running down."

ALTERNATE NAMES: curl-fin sole, curlfin turbot, California turbot.

MAXIMUM RECORDED SIZE: 37 cm (14.5 inches).

DISTRIBUTION: Bahia San Quintin, northern Baja California, Mexico, to Prince William Sound, Gulf of Alaska.

Not well enough known in the Pacific Northwest to stimulate angler interest, the smallish curlfin sole only occasionally ends up in the sportsman's catch as an incidental. While brittle stars and nudibranchs are listed among its prey, other dietary items affix much more readily to a small hook: marine worms, clam siphons and shrimps.

Because the almost-round curlfin sole dwells primarily at depths below 20 m (66 feet) and does not seem to be abundant anywhere in the Pacific Northwest, it seldom crosses the path of the local sport diver.

Seldom harvested by Pacific Northwest trawlers, the thick-bodied curlfin sole, along with several other very similar species, is moderately important in California where trawl nets dragged over level, soft substrates between 38 and 530 m (125 and 1,750 feet) sweep it up for marketing with the other similar species as "turbot." Spawning season for the curlfin sole is from April to August when each ripe adult female lays tiny floating eggs that will hatch into 1.4 mm larvae within a week.

Beachcombers, pierside naturalists and jetty-bound observers never see the deep-dwelling curlfin sole.

Dip a filleted curlfin sole in butter; place it in a greased baking pan, cover the fish with onion, herbs, bread crumbs and a dash of vermouth. Then bake. In some prestigious Californian restaurants the highly regarded curlfin sole appears whole, minus head and entrails.

214

SELECT BIBLIOGRAPHY

During preparation of this book, a very large number of references were consulted for the sake of accuracy, completeness and currency. Unfortunately, space is unavailable to cite all of these excellent scientific works. However, the following short list is presented as a selected group of publications that may assist those readers interested in further study.

Childerhose, R.J. and Trim, M. *Pacific Salmon and Steelhead Trout.* Vancouver: Douglas & McIntyre, 1979.

Eschmeyer, W.N., E.S. Herald, and H. Hammann. *A Field Guide to the Pacific Coast Fishes of North America.* Boston: Houghton Mifflin, 1983.

Forester, J.E. and A.D. Forester. *Fishing: British Columbia's Commercial Fishing History.* Vancouver: Hancock House, 1976.

Hart, J.L. *Pacific Fishes of Canada.* Ottawa: Fisheries Research Board of Canada, Bulletin 180, 1974.

Haw, F. and R.M. Buckley. *Saltwater Fishing in Washington.* Seattle: Stan Jones Publishing Company, 1973.

Jackson, A. *Vancouver Aquarium Seafood Recipes.* Vancouver: Gordon Soules Book Publishers, 1977.

Migadalski, E.C. *How to Make Fish Mounts and Other Fish Trophies.* New York: Ronald Press, 1960.

Nelson, J. *Fishes of the World.* New York: John Wiley and Sons, 1976.

Peden, A.E. *Collecting and Preserving Fishes.* Victoria: Queen's Printer, British Columbia Provincial Museum, Museum Methods Manual 3, 1976.

Starchild, A. and J. Holahan. *Starchild's and Houlahan's Seafood Cookbook.* Seattle: Pacific Press, 1978.

Stewart, H. *Indian Fishing: Early Methods on the Northwest Coast.* Seattle: University of Washington Press, 1977.

Wilson, D. and F. Vander Werff. *New Techniques for Catching Bottom Fish.* Mercer Island, Washington: The Writing Works, 1977.

APPENDIX

Common and scientific names of fish mentioned in the text but not receiving species accounts of their own:

216

217

INDEX

224